Transcriptional Regulation of Flesh Fruit Development and Ripening

Transcriptional Regulation of Flesh Fruit Development and Ripening

Yunbo Luo
China Agricultural University
Beijing, China

WILEY

This edition first published 2023

Registered Offices
John Wiley & Sons, Inc., 111 River Street, Hoboken, NJ 07030, USA
John Wiley & Sons Ltd, The Atrium, Southern Gate, Chichester, West Sussex, PO19 8SQ, UK

For details of our global editorial offices, customer services, and more information about Wiley products visit us at www.wiley.com.

Wiley also publishes its books in a variety of electronic formats and by print-on-demand. Some content that appears in standard print versions of this book may not be available in other formats.

A catalogue record for this book is available from the Library of Congress

Hardback ISBN: 9781394187676; ePub ISBN: 9781394187706; ePDF ISBN: 9781394187683; oBook ISBN: 9781394187690

Cover Image: © Jeffrey Coolidge/Getty Images
Cover Design: Wiley

Set in 9.5/12pt STIX TwoText by Integra Software Services Pvt. Ltd, Pondicherry, India

Contents

Preface

Since the 1990s, I have led the laboratory to engage in the research of postharvest biotechnology and molecular biology of fruits and vegetables in China for the first time. Aiming at the major industrial needs of postharvest quality preservation of fruits, significant advances and contribution have been achieved focusing on the important scientific issue of fruit ripening and senescence regulation, provide a new theoretical basis and way for regulating fruit ripening and senescence and improving fruit quality. With years of research accumulation, we have made many achievements in this field worthy of exchange. We have kept close exchanges with many well-known universities and research institutes at home and abroad through academic exchanges, publishing papers and other ways. Our research has driven the development of this discipline in China like a spark, and we are pleased to see that many laboratories have developed, making China an important force in this field.

The process of fruit development and ripening is a critical stage for metabolism and nutrient formation. Fruit development, ripening, and quality formation are the result of a comprehensive regulation of many transcription factors (TFs). Nearly 60 TFs have been reported to play roles in fruit development, ripening, and quality formation in many plant species, including tomato, apple, pear, peach, grape, banana, strawberry, pepper, and Arabidopsis. According to the different transcription factor families that regulate fruit development and ripening, this monograph is comprised of the following eight parts: Overview of the transcriptional regulation of flesh fruit development and ripening (Chap. 1), Screening method for the identification and characterization of transcription factors regulating flesh fruit development and ripening (Chap. 2), MADS-box transcription factors necessary for flesh fruit development and ripening (Chap. 3), NAC transcription factor family regulation of flesh fruit development and ripening (Chap.4), Role of ERF transcription factors in flesh fruit development and ripening (Chap.5), The ARF side of the fruit tuning of flesh fruit development and ripening (Chap.6), HD-ZIPs are involved in flesh fruit development and ripening (Chap.7), SBP-box transcription factors and flesh fruit development and ripening (Chap.8). Each chapter explains comprehensively analyzing the mechanism of transcriptional regulation of fruit ripening and senescence and investigating the functions and relationships of these TFs have great value in delaying the process of fruit ripening and senescence, effectively maintaining fruit postharvest quality, and extending the storage period.

This book is the first monograph on transcriptional regulation of fruit, fleshy fruit development, ripening and maturity. It embodies the research accumulation and efforts of our research team in the past 30 years and also introduced the globe research progress and achievements in this field in recent years. The content classification of this monograph is very easy for readers in different fields to read. The completion of this book should be grateful to all participants in our group for their hard work on organizing documents, drawings, proofreading, and modification. The book was fulfilled through the joint efforts of all contributors: Daqi Fu, Benzhong Zhu, Hongliang Zhu, Guiqin Qu, Huiqin Tian, Jinlan Xuhui, Shan Li, Yin Gao, Dongyan Cao, Yongfang Yang, Tian Wang, Tongtong Yu, Lanhuan Meng, Yiyang Chu, Cuicui Wang, Ran Li, Xindi Li, Ke Cheng, Peiwen Wu, Lanting Xiang, Shuai Zhang, Liqun Ma. I am very grateful for their suggestions and kind help in pre-paring the manuscripts. Of course, shortcomings in the writing process are unavoidable, and we are eager for reader's criticism to make the book better. I believe that a continuous process of understanding nature from mature aging control to quality formation after harvest. However, the research work is endless, and colleagues have a long way to go.

December 2022

Yunbo Luo
Beijing, China

1

Overview of the Transcriptional Regulation of Flesh Fruit Development and Ripening

1.1 Introduction

As an indispensable part of the human diet, fruit has become increasingly important in daily life. The process of fruit development and ripening is a critical stage for metabolism and nutrient formation. As a result, fruit development, ripening, and the regulation of fruit quality have always been a research hotspot. Fruit development and ripening is a complex but highly regulated physiological and biochemical process; it involves changes in fruit quality such as color, flavor, fragrance, and texture, all of which directly affect the commercial value, postharvest shelf life, and market competitiveness. Fruit development, ripening, and quality formation are the result of a comprehensive regulation of many transcription factors (TFs). Nearly 60 TFs have been reported to play roles in fruit development, ripening, and quality formation in many plant species, including tomato, apple, pear, peach, grape, banana, strawberry, pepper, and *Arabidopsis*. This book focuses on the major transcription factor families that regulate fruit development and ripening, including the MADS-box family, NAC family, ethylene response factor (ERF) family, ARF family, SBP family, and HD-Zip family. In addition, this book introduces methods for the screening and identification of fruit-ripening-related TFs in order to provide theoretical references for the genetic improvement of fruit quality.

1.2 TFs Regulate Fruit Development and Ripening

1.2.1 Overview

The earliest evidence of TFs being involved in fruit ripening and senescence was the discovery of tomato mutants *ripening inhibitor* (*rin*), *nonripening* (*nor*), and *colorless nonripening* (*cnr*). The fruit of these mutants fail to ripen normally;

Transcriptional Regulation of Flesh Fruit Development and Ripening, First Edition. Yunbo Luo.

during the fruit-ripening process, they can neither synthesize large amounts of ethylene nor respond to external ethylene signal (Vrebalov et al. 2002; Manning et al. 2006; Giovannoni 2007; Gapper et al. 2013; Seymour et al. 2013b; Pesaresi et al. 2014). Studies have further shown that the *rin*, *cnr*, and *nor* mutants arose from mutations in MADS-box, SBP-box, and NAC TFs, respectively (Vrebalov et al. 2002; Manning et al. 2006; Giovannoni 2007). In recent years, with the development of genomics, proteomics, transcriptomics, and metabolomics, as well as technologies in DNA–protein and protein–protein interactions, many TFs related to ripening and senescence have been subsequently isolated and identified in tomato, kiwifruit, strawberry, grape, peach, banana, and longan, including EIN3/EIL, AP2/ERF, MADS-box, NAC, and SBP/SPL. Thus, comprehensively analyzing the mechanism of transcriptional regulation of fruit ripening and senescence and investigating the functions and relationships of these TFs have great value in delaying the process of fruit ripening and senescence, effectively maintaining fruit postharvest quality, and extending the storage period.

1.2.2 Model Plant Species for Studying the Transcriptional Regulation of Fruit Development and Ripening

With a long history of cultivation, tomato (*Solanum lycopersicum*) has a small genome (950 Mb) and a short growth cycle (approximately 40 days from fruit set to ripening). Additionally, tomato is distantly related to model plant species such as *Arabidopsis*, rice, and maize. As a result, it has become a model plant species for studying fruit development and ripening in fleshy fruit. A large number of tomato germplasm resources and mutant libraries are currently available (https://solgenomics.net/projects/tomato100), as are high-density genetic map and expressed sequence tag (EST) resources (https://solgenomics.net). Moreover, transient expression and stable transformation technologies are mature in tomato. In addition, the high-quality whole-genome sequence analysis of domesticated tomato was completed in May 2012 (The Tomato Genome Consortium 2012). This information has tremendously promoted the research on tomato functional genomics and molecular genetics. Fruit ripening involves massive regulation of metabolic pathways and significant changes in physiological and biochemical properties. The ripening process requires precise transcriptional regulation of the temporal expression of a series of genes. Thus, studying the functions of ripening-related TFs in tomato has become a research hotspot in plant biology. Bioinformatic analysis has predicted that there are at least 62 TF families in tomato and that they play important regulatory roles in plant hormone biosynthesis, cell expansion, cell wall metabolism, regulation of fruit ripening, and pigment accumulation (Zhang et al. 2011). Investigating the functions of these TFs and their relationships is the foundation of systemically studying tomato fruit ripening and senescence and its regulatory mechanism; this information will provide theoretical reference for targeted tomato fruit quality improvement.

During the early developmental stages of tomato (from fruit set to breaker stage), 11 TF families, including AP2-EREBP, AUX/IAA, GBP, bHLH, HB, ARF, LIM, SPB, HSF, FHA, and PBF-2-Like, are active. At the fruit-ripening stage (breaker stage to fully ripened), seven TF families, including VBI3-VP1, MADS, C2H2, CCAAT, HMG, G2-Like, and MYB-related, have high expression levels. During the process of fruit development and ripening, five TF families, including GRAS, MBF1, MYB, bZIP, and WRKY, have stable expression levels. Among them, the WRKY family exhibited two distinct expression patterns: some WRKY family members exhibited significantly increased expression after the breaker stage, while some other WRKY family members gradually decreased before the breaker stage and gradually increased during fruit ripening (Rohrmann et al. 2011, 2012; Mori et al. 2013). TF families such as C3H, PHD, SET, and Trihelix exhibited very low expression in tomato fruit and may not significantly affect fruit development and ripening. Additionally, many TF families, including ABI3-VP1, bZIP, bHLH, and C2C2-Dof, are expressed in non-fruit tissues, such as leaves, stems, and flowers (Rohrmann et al. 2012).

1.2.3 TF Families that Regulate Fruit Development and Ripening

This book mainly introduces the transcriptional regulation of fruit development and ripening by major TF families, including MADS-box, NAC, ERF, ARF, SBP, and HD-ZIP.

1.2.3.1 MADS-box Family Regulates Fruit Ripening

Members of the MADS-box family are the most extensively studied TFs that play important roles in regulating fruit ripening and senescence. In 2002, Vrebalov et al. found that the tomato *rin* mutants could not ripen normally due to a mutation in a gene encoding a MADS-box TF. Positional cloning revealed that the genomic DNA of the *rin* mutants had a 3 kb deletion, which led to a fusion of two neighboring MADS-box TF genes (*SlMADS-RIN* and *SlMADS-MC*), thereby losing its function (Vrebalov et al. 2002). A subsequent study showed that SlMADS-RIN could bind to the promoter of the ethylene biosynthetic gene *SlACS2* (Ito et al. 2008), indicating that the SlMADS-RIN TF is upstream of ethylene signaling transduction and regulates ethylene biosynthesis. However, the regulatory role of SlMADS-RIN in fruit ripening and senescence is not only through its effect on ethylene signaling, it also acts on other downstream genes. A large number of comparative studies on SlMADS-RIN downstream target genes using proteomics, chromatin immunoprecipitation (ChIP), and electrophoretic mobility shift assay (EMSA) further revealed the transcriptional regulatory network of SlMADS-RIN.

To date, it has been found that SlMADS-RIN downstream target genes include two main categories: ripening-related genes (a, ethylene biosynthetic and responsive genes, such as *SlACS2*, *SlACS4*, *NR*, and *E8*; b, genes encoding cell wall-modifying enzymes, such as *PG*, *TBG4*, and *SlEXP1*; c, genes encoding enzymes for carotenoids metabolism, such as *PSY*; d, genes related to the metabolism of aromatic substances, such as *LOX*, *ADH2*, and *HPL*; e, glycolysis-related genes, such as *PGK*) and TFs (including *RIN* itself, *CNR*, and *SlHB1*) (Fujisawa et al. 2011; Martel et al. 2011; Qin et al. 2012). Fujisawa et al. identified a large number of RIN downstream target genes using transcriptomics and ChIP-chip technologies and found them to be involved in more than 60 biological pathways related to fruit ripening (Fujisawa et al. 2012, 2013). Recently, Wang et al. combined quantitative nuclear proteome analysis with ChIP and EMSA and confirmed that RIN could directly bind to the promoter regions of *SlUBC32* and *PSMB2A*, two genes encoding ubiquitin E2 enzymes, to regulate their expression. Moreover, RIN could affect the level of ubiquitination of many nuclear proteins in the fruit (Wang et al. 2014). RIN can also directly bind to the promoter of *MIR172a* to regulate microRNA accumulation in tomato fruit, thereby further affecting ripening (Gao et al. 2015). In addition to SlMADS-RIN, other tomato MADS TFs, such as TAGL1 (Itkin et al. 2009; Vrebalov et al. 2009), TDR4/FUL1 (Bemer et al. 2012), and MBP7/FUL2 (Bemer et al. 2012; Shima et al. 2013), also play roles in fruit ripening, where TAGL1 is required for normal tomato fruit ripening (Itkin et al. 2009; Vrebalov et al. 2009). Fujisawa et al. identified 860 FUL1 and FUL2 target genes by ChIP-chip, among which 262 were also the target genes of RIN (Fujisawa et al. 2014). It is worth noting that tomato SlMADS1 is a negative regulator of fruit ripening; inhibiting *SlMADS1* expression reduced the ripening time by 2–4 days, significantly increased the carotenoid content in tomato fruit, and significantly upregulated the expression of ethylene biosynthesis-related genes *ACS1A*, *ACS2*, *ACS6*, and *ACO1*. Additionally, the expression of ethylene-responsive genes E4 and E8 was significantly increased, and there was a protein interaction between SlMADS1 and RIN (Dong et al. 2013). Another MADS-box TF SlFYFL is also a negative regulator of ripening. Overexpressing *SlFYFL* resulted in delayed fruit ripening, decreased carotenoid levels, inhibited ethylene biosynthesis, and repressed expression of genes related to ethylene biosynthesis. Furthermore, SlFYFL interacts with SlMADS1 and RIN (Xie et al. 2014). The results of these studies greatly enriched the transcriptional regulatory network of tomato fruit ripening and senescence. Moreover, other studies have shown that the MADS-box gene family is also involved in regulating various important physiological processes, including regulation of seed formation, vegetative organ development, fruit ripening, flower and fruit abscission, root development, and seed embryonic development (Kitazawa Y et al. 2022; Zhang et al. 2022). In addition, MADS-box TFs have been found to be related to fruit ripening and senescence in other plant species, such as grape (Boss et al. 2001), strawberry (Seymour et al. 2011), banana

(Liu et al. 2009a; Elitzur et al. 2010), peach (Tadiello et al. 2009), and bilberry (Jaakola et al. 2010). It is worth noting that in banana, MADS-box genes have been confirmed to play roles in fruit ripening (Liu et al. 2009a; Elitzur et al. 2010). Among these genes, MuMADS1 interacts with ubiquitin-activating enzyme E1 MuUBA (Liu et al. 2013a). MA-MADS5 could also directly regulate ripening-related genes, such as *MA-SPS, MA-ACS1, MA-ACO1, MA-EXP1*, and *MA-lec* (Roy et al. 2012), suggesting that the function of MADS-box TFs in regulating fruit ripening in other plant species (both dicot and monocot) is conserved. Interestingly, although *MaMADS2* is highly homologous to the tomato *RIN* gene, overexpressing *MaMADS2* in tomato *rin* mutants could not restore mutant phenotypes (Elitzur et al. 2010), indicating that differences exist in the regulation mechanism of fruit ripening by MADS-box TFs in different plant species.

1.2.3.2 NAC Family Regulates Fruit Ripening

NAC TFs are plant-specific TFs that have highly conserved sequences consisting of 160-amino acid at the N-terminus, which can be divided into five subdomains (A–E) and may be responsible for the binding of DNA and other proteins (Ernst et al. 2004). The C-terminus exhibits transcriptional activity and is responsible for the activation or repression of their target genes (Ernst et al. 2004). Through binding to the core CGT[G/A] or GCTT on the NAC recognition sequence (NACRS) or NAC-binding sequence (NACBS) in promoters, NAC TFs transcriptionally regulate their target genes and further regulate various biological processes (Lindemose et al. 2014), including fruit ripening and senescence, as well as changes in fruit color, aroma, and texture during this process. The tomato *nonripening* (*nor*) mutants fail to ripen due to a mutation in a NAC domain gene (Martel et al. 2011). Two other NAC TFs related to fruit ripening have been discovered recently in tomato: SlNAC1 (Ma et al. 2014) and SlNAC4 (Ma et al. 2014). In tomato fruit overexpressing *SlNAC1*, the expression of carotenoids and ethylene biosynthetic genes was repressed, causing fruit ripening to be inhibited. However, the pericarp of the fruit was thinner and the level of abscisic acid (ABA) was increased, leading to earlier fruit softening. This result indicates that SlNAC1 affects fruit ripening through an ethylene and ABA-dependent signaling pathway (Ma et al. 2014). In contrast, SlNAC4 is a positive regulator of fruit ripening; inhibiting *SlNAC4* expression resulted in delayed fruit ripening, reduced synthesis of ethylene and carotenoids, and repressed expression of genes related to chlorophyll degradation and fruit ripening. However, the target genes directly regulated by SlNAC4 are still unknown (Zhu et al. 2014). In *Citrus sinensis, CitNAC* expression was detected in the fruit during fruit ripening and senescence, suggesting that fruit ripening and senescence in *Citrus sinensis* may be regulated by CitNAC (Liu et al. 2009b). Recently, Gao et al. reassessed the function of the nor natural mutants. The fact is that NOR is a key transcription factor regulating tomato fruit ripening, involving in the regulation of ethylene biosynthesis, carotenoid accumulation, fruit softening and flavor volatiles metabolism of tomato fruit (Gao et al. 2020, 2022). In addition, NOR could bind to and activate the promoter of SlDML2 (DNA demethylase), affect the promoter methylation level of flavor volatiles biosynthesis genes, and regulate tomato fruit flavor volatiles accumulation via interacting with SlDML2 (Gao et al. 2022). Other transcription factors of NAC family in tomato, SlNAM1/2/3, also participate in tomato fruit ripening. As positive regulators of tomato fruit ripening, they could directly bind to and activate the promoters of SlACS2 and SlACS4, as a result, regulate ethylene production (Gao et al. 2021; Lin et al. 2022). In the fruit of *Actinidia arguta*, AaNAC1/2/3/4 could directly bind to the promoter of *TERPENE SYNTHASE 1* (*AaTPS1*), whereas it could not bind to the promoter of *AcTPS1* in *Actinidia chinensis* fruit (a mutation was present in the NAC-binding element of its promoter). As a result, the aroma profiles of these two types of kiwifruit were different (Nieuwenhuizen et al. 2015). In peach, a NAC TF, Designated Blood (BL), is tightly associated with fruit color (Zhou et al. 2015). DkNAC1/3/5/6 in persimmon is also related to fruit deastringency (Min et al. 2015). EjNAC1 directly regulates the expression of EjPAL1 and Ej4CL1 and functions in the lignification of loquat fruit (Xu et al. 2015). Among the six NAC TFs in banana, MaNAC1–MaNAC6, the expression of *MaNAC1* and *MaNAC2* was significantly induced during the fruit-ripening process and the promoter of *MaNAC2* is activated by external ethylene (Shan et al. 2012). Further studies revealed that MaNAC1 and MaNAC2 can directly bind to the promoter of *MaGRL*, a negative regulator of ethylene signaling, to inhibit its activity. These results indicate that NAC TFs are involved in the regulation of fruit ripening and senescence by influencing ethylene signaling. Studies in rose flowers have found that RhNAC2 could bind to the promoter of a cell wall-modifying gene *RhEXP4A*; RhNAC3 could bind to the promoters of five stress-related genes including *SnRK* (RU09156), *PP2C* (RU23063), *TIP* (RU01501), *GST* (RU07111), and *P5CS* (RU20896); RhNAC100 could bind to the promoter of one cellulose synthase gene, *RhCesA2*, as well as two aquaporin genes, *RhPIP1.1* and *RhPIP2.1*, to regulate postharvest dehydration and senescence in rose flower (Dai et al. 2012; Pei et al. 2013; Jiang et al. 2014a, 2014b). Thus, further investigating and identifying downstream target genes of fruit-ripening-related NAC TFs will deepen the understanding of the regulatory mechanism of NAC TFs in fruit ripening and senescence.

1.2.3.3 ERF Family Regulates Fruit Ripening

The ERF subfamily mainly includes two classes—DREB and ERF—among which ERFs are involved in fruit ripening and senescence and have been in the spotlight. In the earliest report, there were five ERF members in tomato: SlERF1–SlERF4 and SlERF3b. SlERF1 and SlERF4 are transcriptional activators, while SlERF3 is a transcriptional inhibitor (Tournier et al. 2003). *SlERF1* overexpression led to ethylene triple response in etiolated seedlings of transgenic tomatoes, while antisense *SlERF1* resulted in longer shelf life (Li et al. 2007). The expression level of *SlERF2* increases during fruit ripening and senescence (Tournier et al. 2003). *SlERF2* expression is also induced by ethylene; it binds to the DRE element on the *SlACO3* promoter to transcriptionally regulate the feedback loop of ethylene production (Zhang et al. 2009). Using RNAi to inhibit the expression of *SlERF6* could increase the content of carotenoids in tomato fruit and induce the expression of ethylene biosynthesis-related genes *ACS2*, *ACO1*, and *ACO3*. This finding indicates that SlERF6, similar to SlAP2a, could directly or indirectly inhibit the accumulation of carotenoids and ethylene biosynthesis and is a negative regulator of fruit ripening (Li et al. 2012). Recent studies have found that tomato *SlERF.B3* is a multifunctional regulatory gene (Liu et al. 2013b, 2014); the fruit phenotype of the transgenic tomato with the chimeric dominant repressor version (ERF.B3-SRDX) exhibited many changes, including smaller fruit, delayed ripening, enhanced ethylene production, accelerated softening, decreased level of lycopene, and increased level of β-carotene, which resulted in an orange color of the fruit. Correspondingly, fruit of the transgenic tomato with ERF.B3-SRDX had higher expression levels of ethylene biosynthetic genes and genes related to fruit ripening and softening, while genes related to carotenoid biosynthesis were downregulated. In addition, other ripening regulators exhibited significantly different expression levels, including *RIN*, *NOR*, *CNR*, *HB-1*, and *ERF* members (Liu et al. 2014). Deng et al. (2022) demonstrated that SlERF.F12-mediated transcriptional repression of key ripening-related genes ACS2, ACS4, PL, is dependent on the presence of its C-terminal EAR motif (Deng et al. 2022). ERF TFs have been reported to play roles in fruit ripening and senescence in apple, plum, kiwifruit, papaya, persimmon, and banana (Wang et al. 2007; El-Sharkawy et al. 2009; Yin et al. 2010, 2012; Fabi et al. 2012; Min et al. 2012, 2014; Li et al. 2013; Xiao et al. 2013; An et al. 2020). Kiwifruit ERF TFs can directly bind to the promoters of fruit-ripening-related genes to transcriptionally regulate ripening and senescence. AdERF9, which contains the EAR repressor domain, can bind to the promoter of cell wall metabolism-related gene *AdXET5*, which does not contain a GCC box, and inhibit *AdXET5* transcription. This activity suggests that ERFs can also bind to certain unknown *cis*-acting elements (Yin et al. 2010). There are at least 20 deastringency-responsive *DkERF* genes in persimmon fruit, including *DkERF1*, *6*, *9*, *10*, *19*, and *22* (Min et al. 2012, 2014; Yin et al. 2012). Further study has found that DkERF10 and DkERF22 could regulate the promoters of deastringency-related genes *DkADH1* and *DkPDC2*, respectively, while DkERF9 and DkERF10 recognize the promoter of *DkPDC2*, indicating differences among the target genes transcriptionally regulated by DkERF during the process of deastringency (Min et al. 2012, 2014). Two ERF genes in banana (MaERF9 and MaERF11) are closely associated with fruit ripening. MaERF9, as a transcriptional activator, can activate the promoter of ethylene biosynthesis-related gene *MaACO1*, while MaERF11, as a transcriptional repressor, can inhibit the promoter activity of ethylene biosynthesis-related genes *MaACO1* and *MaACS1*. These results showed that ERF TFs control ethylene production via feedback regulation on ethylene biosynthesis-related gene expression, thereby playing a role in regulating fruit ripening (Xiao et al. 2013). The above results indicate that different ERF members may have different regulatory mechanisms during fruit ripening and senescence.

1.2.3.4 ARF Family Regulates Fruit Ripening

Auxin is an important plant hormone involved in plant growth and development. The auxin response factor (ARF) family participates in auxin signal transduction.

ARF proteins consist of three functional domains: a conserved N-terminal DNA-binding domain, a nonconserved middle region with transcriptional activator or repressor activity, and a conserved C-terminal dimerization domain. The expression pattern of ARFs has shown that ARFs are expressed in different organs and different developmental stages in various species. Research in the past 20 years has revealed that ARFs function throughout plant growth and development and are involved in the transcriptional regulation of numerous biological processes related to growth and development. These processes include embryo development, leaf expansion and senescence, lateral root growth, floral organ development and senescence, fruit development, seed development, and responses to external stresses. Many studies have investigated the function of the ARF family in fruit ripening. For example, SlARF4 negatively regulates starch content during tomato fruit ripening (Sagar et al. 2013). SlARF7 negatively regulates the process of fruit set in tomato (De Jong et al. 2011). SlARF9 regulates early fruit development in tomato by negatively regulating cell division (De Jong et al. 2015). SlARF4 participates in the molecular mechanism of auxin and abscisic acid antagonistic regulation of ascorbic acid (AsA) production in tomato (Xu et al. 2022).

1.2.3.5 SBP Family Regulates Fruit Ripening

The SBP gene family is a TF family unique to plants. TFs from the SBP gene family all contain a conserved SBP domain consisting of 76 amino acid residues (SBP-DBD). SBP-DBD has three antiparallel β-sheets at the N-terminus and one or two short helices at the C-terminus. The SBP gene family is broadly involved in the regulation of plant metabolism, growth, and development as well as responses to environmental stresses, such as hormonal responses, the development of fruit, flower, and spores, and defense against pathogen infections. One SBP gene, *LeSPL-CNR*, encodes a key TF that regulates tomato fruit ripening; mutants from DNA methylation in its promoter region significantly inhibited fruit ripening. The fruit of the tomato *cnr* mutants failed to ripen; the pericarp was yellow, and the fruit retained a hard texture even at later stages. The expression of genes encoding many ripening-related enzymes was affected. For example, the synthesis of carotenoids in the pericarp was decreased (Fraser et al. 2001). Compared to wild-type fruit at the same stage, the mutants exhibited a lower level and lower activity of polygalacturonase and pectinesterase (Eriksson et al. 2004).To further identify the

CNR function with CRISPR/Cas 9 in tomato fruit ripening, Gao et al. got the homozygous CRISPR/Cas9-*cnr* mutants with Cas9-free after three generation planting. The phenotype of *CNR* knockout fruits is completely different from that of *cnr* mutants. CRISPR-cnr fruits can turn red, while Cnr mutant ripening fruits appear yellow. Ripening-related genes expression profile seems also different in Cnr mutant and CRISPR-cnr ripen fruits by RNA-Seq. The above results from CRISPR-cnr fruit indicated that the function of CNR in tomato fruit-ripening control still need to be further validated (Gao et al. 2019).

1.2.3.6 HD-ZIP Family Regulates Fruit Ripening

Homeodomain leucine zipper (HD-ZIP) TFs are unique to plants and widely present in most plant species, from moss to higher plants. Based on sequence homology, gene structure, physiological function, and other characteristics, the HD-ZIP TF family can be divided into four classes (I, II, III, IV) (Hu et al. 2012). HD- ZIP Class I proteins are involved in regulating fruit ripening. Class II HD-ZIP proteins are involved in regulating flowering time. Class III HD-ZIP members mainly regulate the development of apical meristems, vascular tissue, and adaxial portions of lateral organs. Class IV HD-ZIP genes are expressed specifically in plant epidermal cells and mainly regulate epidermal cell differentiation, root growth, anthocyanin accumulation, and trichome formation (Fang et al. 2021; Zhang et al. 2021) For example, LeHB-1, a member of the HD-ZIP family, can interact with the promoter of *LeACO1* (an ACC oxidase gene expressed during ripening). Reduced expression of *LeHB-1* led to inhibited ripening, while increasing its expression resulted in altered floral organ morphology in tomato, including the production of multiple flowers within one sepal whorl, fusion of sepals and petals, and the conversion of sepals into carpel-like structures.

1.2.4 Relationships among TF Families

A series of precise transcriptional regulations are present during the tomato fruit-ripening process. Many genes with important biological functions are usually regulated by multiple TFs or regulated at different levels. Furthermore, TFs can self-regulate or regulate each other at the transcriptional level. Rohrmann et al. conducted a correlation analysis of the expression of 134 TFs in fruit (tomato-*Arabidopsis* cross-species coexpression analysis) using qRT-PCR. Through Pearson's correlation analysis, 91 TFs were found to be present in a complex correlation network (Rohrmann et al. 2012). Martel et al. found that binding of SlMADS-RIN to its target loci required the presence of SlSPL-CNR or the gene product regulated by it (Martel et al. 2011). Karlova et al. showed that SlAP2a was downstream of SlNAC-NOR, SlMADS-RIN, and SlSPL-CNR; SlSPL-CNR could directly bind to the promoter of *SlAP2a*, while SlAP2a could negatively regulate the expression of *SlSPL-CNR*. The interactions among MADS TFs are crucial for their functions (Karlova et al. 2011). Leserberg et al. (2008) and Dong et al. (2013) used yeast two-hybrid experiments and found that SlFUL2 interacted with SlMADS-RIN, while SlMADS1 could interact with SlMADS-RIN. Bemer et al. (2012) and Shima et al. (2013) used bimolecular fluorescence complementation (BiFC) and yeast two-hybrid screening, respectively, and confirmed that both SlFUL1 and SlFUL2 could interact with SlMADS-RIN, although SlFUL1 and SlFUL2 did not interact with each other. Fujisawa et al. (2014) conducted a gel retardation assay and found that SlFUL2, SlMADS-RIN, and SlTAGL1 could form tetramers, while SlFUL1 and SlFUL2 could form complexes with SlTAGL1 separately, with SlFUL2-TAGL1 as the predominating form. The direct evidence described above has provided opportunities for revealing the interacting network of tomato fruit-ripening-related TFs and elucidating their regulation of downstream genes, including ripening-related enzymes and receptor proteins at the level of transcription, translation, protein structure, and functional correlation.

1.3 Methods of Screening and Identifying Ripening-related TFs

The methods of screening and identifying ripening-related TFs mainly involve using bioinformatics to screen candidate genes and identifying gene functions by virus-induced gene silencing (VIGS), generating transgenic plants, and gene editing. VIGS uses virus-carrying target gene fragments to infect plants and induce plant endogenous gene silencing and phenotypic changes. In this way, the function of the target gene can be studied based on phenotypic variations. The advantages of using VIGS include short period, rapid results, and high throughput, while the disadvantages are limited vectors, presence of off-target genes, and difficult-to-control environmental conditions. Plant transgenic technology takes external genes originated from animals, plants, microbes, or any other organism, even artificially synthesized genes and then transforms them into the genome of the receiving plant via biological, physical, or chemical approaches that provide genetic stability in the receiving plant and achieve a stable expression and genetic traits against the new genetic background. This approach has the advantage of obtaining research materials with genetic stability but is limited by the long period required to complete the process. Gene editing refers to the technology that utilizes external DNA elements to perform genome modification at specific loci on the chromosome to obtain organisms that contain genetic changes in the desired genomic sequences. This technology can be highly specific to a certain locus in the genome by using certain nucleases. It has advantages of simple design, fast construction, and low cost, but it has the limitation of including possible off-target genes.

References

An, J.P., Zhang, X.W., Bi, S.Q. et al. (2020). The ERF transcription factor MdERF38 promotes drought stress-induced anthocyanin biosynthesis in apple[J]. *Plant J* 101 (3): 573–589.

Bemer, M., Karlova, R., Ballester, A.R. et al. (2012). The tomato FRUITFULL homologs TDR4/FUL1 and MBP7/FUL2 regulate ethylene-independent aspects of fruit ripening. *Plant Cell* 24: 4437–4451.

Boss, P.K., Vivier, M., Matsumoto, S. et al. (2001). A cDNA from grapevine (Vitis vinifera L.), which shows homology to AGAMOUS and SHATTERPROOF, is not only expressed in flowers but also throughout berry development. *Plant Mol Biol* 45: 541–553.

Dai, F., Zhang, C., Jiang, X. et al. (2012). RhNAC2 and RhEXPA4 are involved in the regulation of dehydration tolerance during the expansion of rose petals. *Plant Physiol* 160: 2064–2082.

De Jong, M., Wolters-Arts, M., Garcia-Martinez, J.L. et al. (2011). The Solanum lycopersicum AUXIN RESPONSE FACTOR 7 (SlARF7) mediates cross-talk between auxin and gibberellin signalling during tomato fruit set and development[J]. *J Exp Bot* 62 (2): 617–626.

De Jong, M., Wolters-Arts, M., Schimmel, B.C.J. et al. (2015). Solanum lycopersicum AUXIN RESPONSE FACTOR 9 regulates cell division activity during early tomato fruit development[J]. *J Exp Bot* 66 (11): 3405–3416.

Deng, H., Chen, Y., Liu, Z. et al. (2022). SlERF. F12 modulates the transition to ripening in tomato fruit by recruiting the co-repressor TOPLESS and histone deacetylases to repress key ripening genes[J]. *The Plant Cell* 34 (4): 1250–1272.

Dong, T., Hu, Z., Deng, L. et al. (2013). A tomato MADS-box transcription factor, SlMADS1, acts as a negative regulator of fruit ripening. *Plant Physiol* 163: 1026–1036.

Elitzur, T., Vrebalov, J., Giovannoni, J.J. et al. (2010). The regulation of MADS-box gene expression during ripening of banana and their regulatory interaction with ethylene. *J Exp Bot* 61: 1523–1535.

El-Sharkawy, I., Sherif, S., Mila, I. et al. (2009). Molecular characterization of seven genes encoding ethylene-responsive transcriptional factors during plum fruit development and ripening. *J Exp Bot* 60: 907–922.

Eriksson, E.M. (2004). Effect of the colorless non-ripening mutation on cell wall biochemistry and gene expression during tomato fruit development and Ripening[J]. *Plant Physiol* 136 (4): 4184–4197.

Ernst, H.A., Olsen, A.N., Larsen, S., and Lo, L.L. (2004). Structure of the conserved domain of ANAC, a member of the NAC family of transcription factors. *EMBO Rep* 5: 297–303.

Fabi, J.P., Seymour, G.B., Graham, N.S. et al. (2012). Analysis of ripening-related gene expression in papaya using an Arabidopsis-based microarray. *BMC Plant Biol* 12: 242.

Fang, J.J., Guo, T.T., Xie, Z.W. et al. (2021). The URL1-ROC5-TPL2 transcriptional repressor complex represses the ACL1 gene to modulate leaf rolling in rice. *Plant Physiol* 185: 1722–1744.

Fraser, P.D., Bramley, P., and Seymour, G.B. (2001). Effect of the Cnr mutation on carotenoid formation during tomato fruit ripening.[J]. *Phytochemistry* 58 (1): 75–79.

Fujisawa, M., Nakano, T., and Ito, Y. (2011). Identification of potential target genes for the tomato fruit-ripening regulator RIN by chromatin immunoprecipitation. *BMC Plant Biol* 11: 30–2011.

Fujisawa, M., Nakano, T., Shima, Y., and Ito, Y. (2013). A large-scale identification of direct targets of the tomato MADS box transcription factor RIPENING INHIBITOR reveals the regulation of fruit ripening. *Plant Cell* 25: 371–386.

Fujisawa, M., Shima, Y., Higuchi, N. et al. (2012). Direct targets of the tomato-ripening regulator RIN identified by transcriptome and chromatin immunoprecipitation analyses. *Planta* 235: 1107–1122.

Fujisawa, M., Shima, Y., Nakagawa, H. et al. (2014). Transcriptional regulation of fruit ripening by tomato FRUITFULL homologs and associated MADS box proteins. *Plant Cell* 26: 89–101.

Gao, C., Ju, Z., Cao, D. et al. (2015). MicroRNA profiling analysis throughout tomato fruit development and ripening reveals potential regulatory role of RIN on microRNAs accumulation. *Plant Biotechnol J* 13: 370–382.

Gao, Y., Fan, Z., Zhang, Q. et al. (2021). A tomato NAC transcription factor, SlNAM1, positively regulates ethylene biosynthesis and the onset of tomato fruit ripening. *Plant J.* 108: 1317–1331.

Gao, Y., Lin, Y., Xu, M. et al. (2022). The role and interaction between transcription factor NAC-NOR and DNA demethylase SlDML2 in the biosynthesis of tomato fruit flavor volatiles. *New Phytol.* 235: 1913–1926.

Gao, Y., Wei, W., Fan, Z. et al. (2020). Re-evaluation of the nor mutation and the role of the NAC-NOR transcription factor in tomato fruit ripening. *J Exp Bot* 71: 3560–3574.

Gao, Y., Zhu, N., Zhu, X.F. et al. (2019). Diversity and redundancy of the ripening regulatory networks revealed by the fruitENCODE and the new CRISPR/Cas9 CNR and NOR mutants. *Hort Res* 6: 39–491.

Gapper, N.E., McQuinn, R.P., and Giovannoni, J.J. (2013). Molecular and genetic regulation of fruit ripening. *Plant Mol Biol* 82: 575–591.

Giovannoni, J.J. (2007). Fruit ripening mutants yield insights into ripening control. *Curr Opin Plant Biol* 10: 283–289.

Hu, R., Chi, X., Chai G. et al. (2012). Genome-wide identification, evolutionary expansion, and expression profile of homeodomain-leucine zipper gene family in poplar (Populus trichocarpa). *PLoS One* 7 (2): e31149.

Itkin, M., Seybold, H., Breitel, D. et al. (2009). TOMATO AGAMOUS-LIKE 1 is a component of the fruit ripening regulatory network. *Plant J* 60: 1081–1095.

Ito, Y., Kitagawa, M., Ihashi, N. et al. (2008). DNA-binding specificity, transcriptional activation potential, and the rin mutation effect for the tomato fruit-ripening regulator RIN. *Plant J* 55: 212–223.

Jaakola, L., Poole, M., Jones, M.O. et al. (2010). A SQUAMOSA MADS box gene involved in the regulation of anthocyanin accumulation in bilberry fruits. *Plant Physiol* 153: 1619–1629.

Jiang, G., Jiang, X., Lu, P. et al. (2014a). The rose (Rosa hybrida) NAC transcription factor 3 gene, RhNAC3, involved in ABA signaling pathway both in rose and Arabidopsis. *PLoS One* 9: e109415.

Jiang, X., Zhang, C., Lu, P. et al. (2014b). RhNAC3, a stress-associated NAC transcription factor, has a role in dehydration tolerance through regulating osmotic stress-related genes in rose petals. *Plant Biotechnol J* 12: 38–48.

Karlova, R., Rosin, F.M., Busscher-Lange, J. et al. (2011). Transcriptome and metabolite profiling show that APETALA2a is a major regulator of tomato fruit ripening. *Plant Cell* 23: 923–941.

Kitazawa, Y., Iwabuchi, N., Maejima, K. et al. (2022). A phytoplasma effector acts as a ubiquitin-like mediator between floral MADS-box proteins and proteasome shuttle proteins. *The Plant Cell* 34 (5): 1709–1723.

Leseberg, C.H., Eissler, C.L., Wang, X. et al. (2008). Interaction study of MADS-domain proteins in tomato. *J Exp Bot* 59: 2253–2265.

Li, X., Zhu, X., Mao, J. et al. (2013). Isolation and characterization of ethylene response factor family genes during development, ethylene regulation and stress treatments in papaya fruit. *Plant Physiol Biochem* 70: 81–92.

Li, Y., Zhu, B., Xu, W. et al. (2007). LeERF1 positively modulated ethylene triple response on etiolated seedling, plant development and fruit ripening and softening in tomato. *Plant Cell Rep* 26: 1999–2008.

Li, Y.Y., Mao, K., Zhao, C. et al. (2012). MdCOP1 ubiquitin E3 ligases interact with MdMYB1 to regulate light-induced anthocyanin biosynthesis and red fruit coloration in apple. *Plant Physiol* 160: 1011–1022.

Lin, D., Zhu, X., Qi, B. et al. (2022). SlMIR164A regulates fruit ripening and quality by controlling SlNAM2 and SlNAM3 in tomato. *Plant Biotechnol J* 20: 1456–1469.

Lindemose, S., Jensen, M.K., Van de Velde, J. et al. (2014). A DNA-binding-site landscape and regulatory network analysis for NAC transcription factors in Arabidopsis thaliana. *Nucleic Acids Res* 42: 7681–7693.

Liu, J., Xu, B., Hu, L. et al. (2009a). Involvement of a banana MADS-box transcription factor gene in ethylene-induced fruit ripening. *Plant Cell Rep* 28: 103–111.

Liu, J.H., Zhang, J., Jia, C.H. et al. (2013a). The interaction of banana MADS-box protein MuMADS1 and ubiquitin-activating enzyme E-MuUBA in post-harvest banana fruit. *Plant Cell Rep* 32: 129–137.

Liu, M., Diretto, G., Pirrello, J. et al. (2014). The chimeric repressor version of an Ethylene Response Factor (ERF) family member, Sl-ERF.B3, shows contrasting effects on tomato fruit ripening. *New Phytol* 203: 206–218.

Liu, M., Pirrello, J., Kesari, R. et al. (2013b). A dominant repressor version of the tomato Sl-ERF.B3 gene confers ethylene hypersensitivity via feedback regulation of ethylene signaling and response components. *Plant J* 76: 406–419.

Liu, Y., Baig, M.N.R., Fan, R. et al. (2009b). Identification and expression pattern of a Novel NAM, ATAF, and CUC-Like Gene from Citrus sinensis Osbeck. *Plant Mol Biol Rep* 27: 292–297.

Ma, N., Feng, H., Meng, X. et al. (2014). Overexpression of tomato SlNAC1 transcription factor alters fruit pigmentation and softening. *BMC Plant Biol* 14: 351.

Manning, K., Tor, M., Poole, M. et al. (2006). A naturally occurring epigenetic mutation in a gene encoding an SBP-box transcription factor inhibits tomato fruit ripening. *Nat Genet* 38: 948–952.

Martel, C., Vrebalov, J., Tafelmeyer, P., and Giovannoni, J.J. (2011). The tomato MADS-box transcription factor RIPENING INHIBITOR interacts with promoters involved in numerous ripening processes in a COLORLESS NONRIPENING-dependent manner. *Plant Physiol* 157: 1568–1579.

Min, T., Fang, F., Ge, H. et al. (2014). Two novel anoxia-induced ethylene response factors that interact with promoters of deastringency-related genes from persimmon. *PLoS One* 9: e97043.

Min, T., Wang, M.M., Wang, H. et al. (2015). Isolation and expression of NAC genes during persimmon fruit postharvest astringency removal. *Int J Mol Sci* 16: 1894–1906.

Min, T., Yin, X.R., Shi, Y.N. et al. (2012). Ethylene-responsive transcription factors interact with promoters of ADH and PDC involved in persimmon (Diospyros kaki) fruit de-astringency. *J Exp Bot* 63: 6393–6405.

Mori, K., Lemaire-Chamley, M., Asamizu, E. et al. (2013). Comparative analysis of common genes involved in early fruit development in tomato and grape. *Plant Biotechnol* 30: 213–295.

Nieuwenhuizen, N.J., Chen, X., Wang, M.Y. et al. (2015). Natural variation in monoterpene synthesis in kiwifruit: transcriptional regulation of terpene synthases by NAC and ETHYLENE-INSENSITIVE3-like transcription factors. *Plant Physiol* 167: 1243–1258.

Pei, H., Ma, N., Tian, J. et al. (2013). An NAC transcription factor controls ethylene-regulated cell expansion in flower petals. *Plant Physiol* 163: 775–791.

Pesaresi, P., Mizzotti, C., Colombo, M., and Masiero, S. (2014). Genetic regulation and structural changes during tomato fruit development and ripening. *Front Plant Sci* 5: 124.

Qin, G., Wang, Y., Cao, B. et al. (2012). Unraveling the regulatory network of the MADS box transcription factor RIN in fruit ripening. *Plant J* 70: 243–255.

Rohrmann, J., McQuinn, R., Giovannoni, J.J. et al. (2012). Tissue specificity and differential expression of transcription factors in tomato provide hints of unique regulatory networks during fruit ripening. *Plant Signal Behav* 7: 1639–1647.

Rohrmann, J., Tohge, T., Alba, R. et al. (2011). Combined transcription factor profiling, microarray analysis and metabolite profiling reveals the transcriptional control of metabolic shifts occurring during tomato fruit development. *Plant J* 68: 999–1013.

Roy, C.S., Roy, S., Nag, A. et al. (2012). Characterization of an AGAMOUS-like MADS box protein, a probable constituent of flowering and fruit ripening regulatory system in banana. *PLoS One* 7: e44361.

Sagar, M., Chervin, C., Mila, I. et al. (2013). SlARF4, an Auxin response factor involved in the control of sugar metabolism during tomato fruit development. *Plant Physiol* 161 (3): 1362–1374.

Seymour, G.B., Chapman, N.H., Chew, B.L., and Rose, J.K. (2013a). Regulation of ripening and opportunities for control in tomato and other fruits. *Plant Biotechnol J* 11: 269–278.

Seymour, G.B., Ostergaard, L., Chapman, N.H. et al. (2013b). Fruit development and ripening. *Annu Rev Plant Biol* 64: 219–241.

Seymour, G.B., Ryder, C.D., Cevik, V. et al. (2011). A SEPALLATA gene is involved in the development and ripening of strawberry (Fragaria x ananassa Duch.) fruit, a non-climacteric tissue. *J Exp Bot* 62: 1179–1188.

Shan, W., Kuang, J.F., Chen, L. et al. (2012). Molecular characterization of banana NAC transcription factors and their interactions with ethylene signalling component EIL during fruit ripening. *J Exp Bot* 63: 5171–5187.

Shima, Y., Kitagawa, M., Fujisawa, M. et al. (2013). Tomato FRUITFULL homologues act in fruit ripening via forming MADS-box transcription factor complexes with RIN. *Plant Molecular Biology* 82: 427–438.

Tadiello, A., Pavanello, A., Zanin, D. et al. (2009). A PLENA-like gene of peach is involved in carpel formation and subsequent transformation into a fleshy fruit. *J Exp Bot* 60: 651–661.

The Tomato Genome Consortium. (2012). The tomato genome sequence provides insights into fleshy fruit.

Tournier, B., Sanchez-Ballesta, M.T., Jones, B. et al. (2003). New members of the tomato ERF family show specific expression pattern and diverse DNA-binding capacity to the GCC box element. *FEBS Lett* 550: 149–154.

Vrebalov, J., Pan, I.L., Arroyo, A.J.M. et al. (2009). Fleshy fruit expansion and ripening are regulated by the tomato SHATTERPROOF gene TAGL1. *Plant Cell* 21: 3041–3062.

Vrebalov, J., Ruezinsky, D., Padmanabhan, V. et al. (2002). A MADS-box gene necessary for fruit ripening at the tomato ripening-inhibitor (rin) locus. *Science* 296: 343–346.

Wang, A., Tan, D., Takahashi, A. et al. (2007). MdERFs, two ethylene-response factors involved in apple fruit ripening. *J Exp Bot* 58: 3743–3748.

Wang, Y., Wang, W., Cai, J. et al. (2014). Tomato nuclear proteome reveals the involvement of specific E2 ubiquitin-conjugating enzymes in fruit ripening. *Genome Biol* 15.

Xiao, Y.Y., Chen, J.Y., Kuang, J.F. et al. (2013). Banana ethylene response factors are involved in fruit ripening through their interactions with ethylene biosynthesis genes. *J Exp Bot* 64: 2499–2510.

Xie, Q., Hu, Z., Zhu, Z. et al. (2014). Overexpression of a novel MADS-box gene SlFYFL delays senescence, fruit ripening and abscission in tomato. *Sci Rep* 4: 4367.

Xu, Q., Wang, W., Zeng, J. et al. (2015). A NAC transcription factor, EjNAC1, affects lignification of loquat fruit by regulating lignin. *Postharvest Biol. Technol.* 102: 25–31.

Xu, X., Zhang, Q., Gao, X. et al. (2022). Auxin and abscisic acid antagonistically regulate ascorbic acid production via the SlMAPK8–SlARF4–SlMYB11 module in tomato. *Plant Cell* 34: 4409–4427.

Yin, X.R., Allan, A.C., Chen, K.S., and Ferguson, I.B. (2010). Kiwifruit EIL and ERF genes involved in regulating fruit ripening. *Plant Physiol* 153: 1280–1292.

Yin, X.R., Shi, Y.N., Min, T. et al. (2012). Expression of ethylene response genes during persimmon fruit astringency removal. *Planta* 235: 895–906.

Zhang, H., Jin, J., Tang, L. et al. (2011). PlantTFDB 2.0: update and improvement of the comprehensive plant transcription factor database. *Nucleic Acids Res* 39: D1114–D1117.

Zhang, S., Yao, J., Wang, L. et al. (2022). Role of grapevine SEPALLATA-related MADS-box gene VvMADS39 in flower and ovule development. *Plant J* 111: 1565–1579.

Zhang, T., You, J., Zhang, Y. et al. (2021). LF1 regulates the lateral organs polarity development in rice. *New Phytol* 231: 1265–1277.

Zhang, Z., Zhang, H., Quan, R. et al. (2009). Transcriptional regulation of the ethylene response factor LeERF2 in the expression of ethylene biosynthesis genes controls ethylene production in tomato and tobacco. *Plant Physiol* 150: 365–377.

Zhou, H., Lin-Wang, K., Wang, H. et al. (2015). Molecular genetics of blood-fleshed peach reveals activation of anthocyanin biosynthesis by NAC transcription factors. *Plant J* 82: 105–121.

Zhu, M., Chen, G., Zhou, S. et al. (2014). A new tomato NAC (NAM/ATAF1/2/CUC2) transcription factor, SlNAC4, functions as a positive regulator of fruit ripening and carotenoid accumulation. *Plant Cell Physiol* 55: 119–135.

2

Screening Method for the Identification and Characterization of Transcription Factors Regulating Flesh Fruit Development and Ripening

2.1 Bioinformatics

Bioinformatics is a subdiscipline of biology and computer science concerned with the storage, retrieval, and analysis of biological data. It is a major field at the frontier of life science and one of the core fields of the natural sciences in the 21st century. The primary research focuses of bioinformatics are genomics and proteomics, in which nucleic acid and protein sequences are analyzed to extract their biological structural and functional information.

2.1.1 Overview

2.1.1.1 Introduction

Bioinformatics is a new field that analyzes genomic DNA sequence data to obtain protein-coding information to simulate and predict protein spatial structures, which enables drug design based on specific protein functions. Genomic informatics, protein spatial structure simulation, and drug design are the three important components of bioinformatics. Hence, bioinformatic research includes the following three main aspects: (1) research on new algorithms and statistical methods; (2) data analysis and interpretation; and (3) development of new tools for effective utilization and management of data.

Bioinformatics is essentially a combination of molecular biology and information technology (especially internet technology). It uses a wide range of biological data as research materials and computing as a research tool and employs various research methods, including biological data searching (e.g., collection and screening), processing (e.g., editing, sorting, management, and displaying), and utilization (e.g., calculation and simulation).

Since the 1990s, with the launching of various genome sequencing projects, the breakthroughs in technologies for determining molecular structures and the popularization of

Transcriptional Regulation of Flesh Fruit Development and Ripening, First Edition. Yunbo Luo.

the internet have prompted the emergence and expansion of hundreds of biological databases. Severe challenges to bioinformatics researchers thus arise, giving rise to a series of questions: What information is carried by these hundreds of millions of ACGT sequences? What is the mechanism behind genomic control of the development of an organism? How have the genomes evolved?

Another challenge in bioinformatics is to predict a protein's structure from its amino acid sequence, which has plagued theoretical biologists for more than half a century. Finding an effective method for predicting protein structure is increasingly urgent. In 1991, Nobel Laureate W. Gilbert stated that "in the current paradigm, the attack on the problems of biology is viewed as being solely experimental.... The new paradigm, now emerging, is that all the genes will be known (in the sense of being resident in databases available electronically), and that the starting point of a biological investigation will be theoretical. An individual scientist will begin with a theoretical conjecture, only then turning to experiment to follow or test that hypothesis."

Bioinformatics research has been following a general trajectory from genomics to proteomics, systems biology, and comparative genomics. In 1989, the International Conference on Computer Modeling of Biochemical Systems Theory and Biomathematics was held in the United States, marking the development of bioinformatics to the era of computational biology and computational systems biology.

To summarize in easier-to-understand language instead of quoting lengthy bioinformatics definitions, the milestone progress of biological genome sequencing projects (including the Human Genome Project) has increased biological data (e.g., data related to birth, aging, and death) at an unprecedented rate of doubling every 14 months. Concurrently, the development of the internet has prompted the booming and expansion of hundreds of biological databases, representing the first developmental stage of a bioinformatics industry that has achieved only the acquisition of raw biological information. Most bioinformatics enterprises at this stage make profits by providing biological databases, a successful representative of which is Guangzhou SALIAI Stemcell Science and Technology Co., Ltd., best known for its human genome sequencing service.

The advanced stage of the bioinformatics industry is reflected by the bioinformatics-centered postgenomic era, during which the typical application is new drug innovation on the basis of bioinformatics.

2.1.1.2 Stages of Development

The pregenomic era (before the 1990s): This stage is represented mainly by the establishment of various sequence comparison algorithms, establishment of biological databases, development of retrieval tools, and DNA and protein sequence analysis, etc.

The genomic era (from the 1990s to 2001): This stage mainly involved large-scale genome sequencing, gene identification and discovery, systematic construction of network databases, and development of interactive interface tools, etc.

The postgenomic era (from 2001 to the present): With the completion of the Human Genome Project and genome sequencing of various model organisms, biological science has entered the postgenomic era, during which the research focus has shifted from genomic structure to gene function. This shift is reflected by the emergence of functional genomics, for which the preliminary genomic research is called structural genomics.

2.1.1.3 Brief Introduction to the Development of Bioinformatics

Bioinformatics is based on molecular biology. Therefore, to understand bioinformatics, it is necessary to first understand the development of molecular biology. Structural and functional research of biological macromolecules in biological cells has a long history. In 1866, based on experimental findings, Mendel hypothesized that genetic factors exist in the form of biological components. In 1871, Miescher isolated DNA from the nuclei of dead white blood cells. However, chromosomal proteins were still thought to carry genes, and DNA played only a minor role until 1944, when Avery and McCarty confirmed DNA as the genetic material of life and organs. In 1944, Chargaff discovered that the amount of guanine (G) is always equal to that of cytosine (C) and the amount of adenine (A) is equal to that of thymine (T) in DNA, the famous Chargaff rules. Furthermore, Wilkins and Franklin determined the structure of DNA fibers by using X-ray diffraction technology. In 1953, James Watson and Francis Crick published their deduction of the three-dimensional (3D) double helical structure of DNA in *Nature*. They stated that DNA forms a double-stranded helix that has backbones made of alternating sugar (deoxyribose) and phosphate groups and that the bases on deoxyribose form base pairs between double-stranded sugar-phosphate chains following Chargaff's rules. This model shows that DNA has a self-complementary structure and that the genetic information stored in DNA can be replicated precisely following the base-pairing rules. Watson and Crick's theories laid the foundation for molecular biology, and their DNA double helix model predicted the rules of DNA replication. In 1956, Kornberg isolated DNA polymerase I from *E. coli*, which can be used to connect four types of dNTPs to produce DNA. DNA replication requires a DNA template. Meselson and Stahl (1958) experimentally revealed that DNA replication is semiconservative. In 1954, Crick outlined the central dogma to describe the flow of genetic information in a biological system and stated that DNA is the template for RNA synthesis, while RNA is the template for protein synthesis. The central dogma has played a particularly important role in guiding the development of molecular biology and bioinformatics. Nirenberg and Matthai (1963) made great strides and deciphered the genetic code for 20 amino acids. Subsequently, the discovery of restriction endonuclease and the development of recombinant DNA cloning technology laid a technical foundation for genetic engineering. Research on molecular biology greatly promoted the development of life science, inevitably leading to the emergence of bioinformatics. In February 2001, the completion of the Human Genome Project brought bioinformatics into being. With the rapid development of automatic DNA sequencing technology, the amount of nucleic acid sequence data in public DNA databases is increasing at the rate of 106 bp per day, generating an immense amount of biological information data. Undoubtedly, as the data accumulation era is turning into a data interpretation era, innumerous data accumulation is bound to increase the potential for breakthrough discoveries. Interdisciplinary "bioinformatics" has developed under this premise. In brief, the core aim of bioinformatics is to deepen the knowledge of DNA sequences, structure, and evolution and their relationships with biological function through statistically computing and analyzing DNA sequences. Bioinformatics research involves many fields, including molecular biology, molecular evolution and structural biology, statistics, and computer science. Bioinformatics is an area of science that has a very broad applicability and includes the acquisition, processing, storage, distribution, and interpretation of genomic information with genomic informatics as its core. The key content of genomic informatics is to "understand" the nucleotide sequence of a genome, i.e., the exact position of all genes on the chromosome and the function of each DNA fragment. After a new gene is identified, the spatial structure of its encoded protein is simulated and predicted for new drug design based on the specific function of the protein. Understanding the regulatory mechanism of gene expression is also an important component of bioinformatics, with the aim of revealing intrinsic patterns of disease for diagnosis and treatment based on the role of biomolecules in gene regulation. The research objective of bioinformatics is to reveal "the structural complexity of genomic information and the fundamental rules of genetic language" and decipher the genetic language of organisms. Bioinformatics has become an important discipline in the development of life science and is at the frontier of life science research.

2.1.1.4 Research Direction

In just over ten years, bioinformatics research has been advanced in various directions of sequence alignment, gene identification, molecular evolution, sequence contig assembly, genetic coding, and biological systems, which are briefly introduced below.

1) Sequence alignment

Sequence alignment is essentially a comparison identifying the similarities or dissimilarities between two or more sequences. To achieve this end, sequence alignment involves the following steps: (1) reconstructing the complete DNA sequence from overlapping sequences; (2) determining the storage systems of physical and genetic map data based on probe data under various experimental conditions; (3) screening and comparing DNA sequences against the database to determine the similarities between two or more sequences; (4) searching for relevant sequences and subsequences in the database to determine continuous production patterns of nucleotides; and (5) identifying the informative components in protein and DNA sequences. Sequence alignment considers the biological traits of DNA sequences, such as insertions and deletions, abbreviated as "indels," as well as substitutions that occur locally in the sequence. The objective function of sequences is to obtain the minimum distance weighted sum or the maximum similarity sum of the mutation sets between the sequences. The alignment methods include global alignment, local alignment, and gap penalty. The dynamic programming algorithm, a commonly used sequence alignment approach, is applicable for short gene sequences. For sequences with a considerable length (e.g., human DNA sequences up to 109 bp), the dynamic programming algorithm and even algorithms with linear complexity are not ideal. This challenge inevitably prompted the introduction of heuristic methods, e.g., the well-known BLAST and FASTA algorithms and their improved versions.

2) Protein alignment

Protein alignment involves the comparison of two or more protein molecules to determine their similarities or dissimilarities in spatial structure. Protein structure and function are closely related; therefore, it is generally believed that proteins with similar functions are similar in structure. A protein is a long amino acid chain with a length ranging from 50 to 1000–3000 amino acids (AAs). Protein functions are highly diverse, e.g., enzymes, substance storage and transportation, signal transmission, and antibodies. The 3D structure of a protein is intrinsically determined by its amino acid sequence, and generally, protein structure consists of four levels. Research on protein structure and prediction holds great benefits for many fields. In medicine, protein structures suggest the biological functions of proteins, thereby facilitating target identification for developing docking drugs; in agriculture, knowledge of protein structures informs genetic engineering to improve crop quality; and it can improve industrial synthesis of enzymes. Direct protein structure alignment is of great value because the protein 3D structure is more stably preserved in evolution than its primary structure and contains more information than its amino acid sequence. The premise of protein 3D structure research is an assumption that an amino acid sequence forms a specific 3D structure (although this is not always the case) according to the minimum energy principle in physics. The structure of an unknown protein can be predicted by homology modeling and threading based on the structural patterns of known proteins identified from observation and summarization. Homology modeling is performed to detect protein structures with high similarity (in which more than 30% amino acids are the same), and threading is performed to compare different protein structures in evolutionary families. However, the current research on protein structure prediction falls far short of meeting the actual needs.

3) Gene identification

Gene identification seeks to correctly identify the range and exact position of genes in a given genome sequence. Although noncoding regions, which consist of introns, are generally discarded after protein synthesis, a gene cannot be replicated experimentally if its noncoding region is removed. This finding indicates that the genetic language embedded in a DNA sequence requires the presence of both coding and noncoding regions. No universal guidance has been developed for analyzing noncoding DNA sequences yet. Not all sequences in the human genome act as templates to encode proteins, and in fact, the coding sequences that have been identified account for only 3–5% of the total human gene sequence. Manual search of such a large genome sequence would be impossible. The methods for coding region detection include measuring codon frequencies, the first- and second-order Markov chains, open reading frames (ORFs), promoter identification, hidden Markov model (HMM), GENSCAN, splice alignment, etc.

4) Molecular evolutionary analysis

Molecular evolution analysis is performed to study the evolution of species and build a phylogenetic tree based on the similarities and differences in the same gene sequence between different species. Alignments of DNA sequences and their encoded amino acid sequences, or even related proteins, can be used to investigate molecular evolution based on the assumption that similar species are genetically similar. The similarities and differences among different species are identified through comparative analysis at the genomic level. Early research often used external traits, such as size, skin color, and number of limbs, as evidence for evolution. With the completion of genome sequencing for more model species, molecular evolution can be explored from a whole-genome perspective. Gene alignment among different species may involve three types of genes: 1) orthologous genes that retain the same function in different species; 2) paralogous genes that have different functions in the same species; and 3) xenologs, which are genes passing between organisms by special means, such as being injected through viral infection. To explore molecular evolution, a common approach is to construct evolutionary trees using feature-based (i.e., a specific base position in a DNA sequence or protein) and distance-based (aligned fractions) methods, as well as traditional clustering methods such as the unweighted pair-group method using arithmetic averages (UPGMA).

5) Sequence contig assembly

The existing sequencing technologies can only detect the sequences with a length of up to approximately 500 bp in each reaction. For example, shotgun sequencing used for human genes requires the generation of contigs from a large number of short fragments. Contig assembly refers to the process in which short fragments are spliced to form longer sequence contigs until a complete sequence is obtained. From the algorithmic level, sequence contig assembly is an NP-complete problem.

6) Genetic coding

It is generally believed that the relationships between codons and amino acids were initiated in fortuitous events during biological evolution, then became fixed in the common ancestor of modern organisms and are preserved to the present. In addition to this "frozen" theory, three different theories were proposed to explain genetic coding from the perspectives of selection/optimization, chemistry, and history. With the completion of more biological genome sequencing projects, an increasing amount of evidence will be provided for research on the origin of genetic code and testing of the abovementioned theories.

7) Biological systems

The development of large-scale experimentation technologies and data accumulation have prompted the rise of systems biology, a new research hotspot in the postgenomic era, which studies and analyzes biological systems at the global and system level and reveals their developmental rules. At present, the research content of systems biology includes the simulation of biological systems (Aigner et al. 2007), system stability analysis (Nikolov et al. 2007), system robustness analysis (Baliga 2008), etc. Modeling languages represented by the Systems Biology Markup Language (SBML) (Saez-Rodriguez et al. 2007) are developing rapidly. A variety of methods, including Boolean network (Chen et al. 2004), differential equation (Soula and Chow 2007), stochastic process (Koschorreck et al. 2007), and discrete-event dynamic systems (Kayser et al. 2007), have been applied in systems analysis. Various models were established by borrowing the modeling methods of circuits and other physical systems, and many studies have attempted to solve the complexity problem of biological systems using macro-analytical concepts such as information flow, entropy, and energy flow (Lijoi et al. 2007). Nevertheless, it will take a long time to establish theoretical models for biological systems because, although the experimental observation data are rapidly accumulating, the data required for model identification of biological systems far exceed the current data output. For example, the number of sampling points required for time-series chip data is beyond the capacity of the traditional time series modeling method. In addition to the main difficulty of high experimental cost for system modeling, system description and modeling methods also require groundbreaking developments.

2.1.1.5 Technical Methods

Bioinformatics research strategies include four steps: biological database establishment, data retrieval, data processing, and data utilization with data (database) as the core. Thus, this process involves more than the simple compilation of biological knowledge or simple application of scientific knowledge in mathematics, physics, and information science. A series of technologies, from machine learning to statistical data analysis and system description, have rapidly advanced to meet the need for handling innumerous amounts of data and complex backgrounds in bioinformatics. It is extremely difficult to use traditional statistical analysis to process immense amounts of data, complex noise patterns, and massive time-varying data; thus, more flexible data analysis approaches, such as nonparametric statistics (Fukuoka et al. 2007) and cluster analysis, are needed. Analysis of high-dimensional data requires feature-space compression technologies such as partial least squares (PLS). The development of computing algorithms must fully consider the time and space complexity of the algorithm and use parallel computing, grid computing, and other technologies to expand the applicability of algorithms.

2.1.1.6 Others

Bioinformatics is a new interdisciplinary field developed in recent years. It comprehensively utilizes biological, computer science, and information technologies to process a large amount of complex biological data and reveal their biological significance, thereby guiding the development of scientific research and agricultural technology.

Gene expression profiling, metabolic network analysis, gene chip design, and proteomics data analysis have gradually become key research fields that newly emerged in bioinformatics. The disciplines derived from bioinformatics include structural genomics, functional genomics, comparative genomics, proteomics, pharmacogenomics, Chinese medicine genomics, tumor genomics, molecular epidemiology, and environmental genomics, which have become important tools in the research of systems biology. From a developmental perspective, it is clear that genetic engineering has entered the postgenomic era. Nevertheless, it is important to have a clear understanding of the potential misleading factors in areas closely related to bioinformatics, such as machine learning and mathematics.

2.1.2 Expression Profile Analysis

Gene expression profiling is conducted to determine the types and abundance of genes expressed in the tissue under specific circumstances or in a specific cell, namely, the gene expression profile, through constructing an unbiased cDNA library, large-scale cDNA sequencing, collecting cDNA sequence fragments, and then qualitatively and quantitatively analyzing mRNA population composition. Differential expression profiling includes gene expression profiling and protein expression profiling.

2.1.2.1 Gene Expression Profile

A gene-chip assay requires two groups of tissue samples (i.e., the experimental and control groups), e.g., normal and tumor tissues, tissues at different stages of development, or cells or tissues before and after drug treatment. During reverse transcription, the mRNA samples from the experimental and control groups are labeled respectively with red and green fluorescence and are then mixed together and hybridized to microarray probes. After a proper elution step, the chip is scanned with a laser scanner to obtain the fluorescence intensity image for both red and green fluorescence. Finally, the red and green fluorescence intensities (Cy5 and Cy3) of each spot on the microarray can be obtained by using special image analysis software, and the Cy5/Cy3 ratio represents the expression level of the gene in the experimental sample.

2.1.2.2 Acquisition of Gene Expression Profile

To obtain a complete gene expression profile, the main steps are as follows:

1) Preparation of a gene chip

A large number of probes are immobilized on a carrier such as a silica gel, a glass plate, a polypropylene, or a nylon membrane in a specific arrangement to form a DNA chip. There are many types of chips, and their preparation methods are not the same. The preparation methods can be divided into two categories: in situ synthesis and direct spotting.

2) Preparation of fluorescently labeled probes

The gene probe to be analyzed is subjected to purification, reverse transcription, and amplification-grade fluorescent labeling before chip hybridization. The most common fluorescent labeling method is to label the control sample with Cy3-dUTP (green fluorescence) and the laboratory sample with Cy5-dUTP (red fluorescence).

3) Hybridization of the labeled probe to the chip

The hybridization conditions are selected, and the prepared fluorescent probe is hybridized with the chip. After a defined time, the gene chip is washed thoroughly to remove the unbound and nonspecific signal and is then ready for scanning of fluorescence microarray.

4) Scanning of the hybridized chip

The hybridized chip is excited with a laser of a specific wavelength at which the probe on the chip emits fluorescence of a different wavelength. Two monochromatic images of Cy5 and Cy3 can be obtained by detecting the fluorescence intensity of the probe using a confocal laser scanner. The two images are pseudocolor processed separately and then superimposed as an image, i.e., the hybrid image, which is the original data obtained by the experiment. The fluorescence intensity value at a spot in the image indirectly gives the relative abundance of the corresponding probe in the gene chip. Green indicates that the control sample has a higher expression level, yellow indicates that the two samples have an equal level of expression, and red indicates that the laboratory sample has a higher expression level. This approach can demonstrate the difference in the relative expression level of a gene between two different samples.

5) The gene expression profile data are extracted from the hybrid image; i.e., the original hybrid image is converted into gene expression profile data.

2.2 Virus-induced Gene Silencing (VIGS)

In this section, VIGS is described in detail, including its principles and commonly used methods. To date, this technology has been widely used to study gene function in plants (Pan et al. 2013b; Li et al. 2022; Rössner et al. 2022).

VIGS is a common plant gene-silencing technique that infects a plant with viral vectors carrying a fragment of a gene of interest to initiate endogenous gene silencing and phenotypic changes in the plant, enabling the study of the function of the target gene based on phenotypic variations. VIGS was developed from the research on virus–host and virus–transgene interactions (Lindbo et al. 1993; Kumagai et al. 1995). After the viral vectors carrying a fragment of a gene of interest are introduced into the plant cells, the cells recognize the viruses and initiate the defense mechanism to kill the viruses and destroy any exogenous genes they carried, causing transcriptional changes and a specific mRNA degradation of the target gene in the plant itself (Ruiz et al. 1998; Waterhouse et al. 2001).

With the development of molecular biotechnology, a variety of methods have emerged to obtain new plant gene information, such as genome sequencing, gene chip, and differential screening technology (Zegzouti et al. 1999; Richmond and Somerville 2000). However, sequencing is only the first step, and functional identification is more important for a full understanding of a gene. In addition, high-throughput gene function identification methods are urgently needed to meet the demand for large-scale gene sequencing. Antisense transgenesis was a common and effective method to study the function of plant genes in the past, but this method required tissue culture to produce transgenic plants (Herrera-Estrella et al. 2004). Due to the unique double-stranded complementary structure of genes, transgenesis can effectively improve the inhibition efficiency of target genes in transgenic plants (Wesley et al. 2001), but it still requires production of transgenic plants. VIGS does not require time-consuming genetic transformation and tissue culture procedures, and therefore, it can serve as a potentially important tool for reverse genetic research of target genes and accelerate gene function analysis. Compared with traditional transgenic technology, VIGS has the following advantages. First, VIGS is simple to perform. It can achieve good results through *Agrobacterium* infection without complex genetic transformation procedures. Second, the VIGS technique is time-saving. It usually achieves a good effect within two to three weeks, while transgenic technology needs one to two years to produce transgenic plants and complete complicated passaging and genetic identification. Third, viral vectors have been improved to allow VIGS to be applied in both monocots and dicots. In contrast, due to a high selectivity of transgene expression in plants, genetic transformation is difficult in monocots. Fourth, VIGS does not show target gene selectivity and thus can be used to study any gene related to plant growth and development. Instead, transgenic technology is not suitable for genes related to early embryonic development because their expression inhibition may lead to the death of the plants. Fifth, VIGS is a better tool for functional analysis of gene members in the same family. Through the design of appropriate target gene fragments, VIGS can specifically silence single or multiple genes to solve the functional redundancy problem in a gene family. Sixth, VIGS can be effectively applied to functionally analyze each gene involved in disease-resistance signal transduction. To date, VIGS has been successfully applied to the research on *N* gene-mediated resistance (Liu et al. 2002c). Rui et al. established a library using potato virus X (PVX) and obtained a new heat shock protein gene, HSP90, by high-throughput VIGS screening of 4992 cDNA clones (Lu 2002). Researchers from Yale University have effectively combined the Gateway cloning technology with a tobacco rattle virus (TRV) vector to create a new vector, which can be used for functional identification of expressed sequence tags (ESTs) (Liu et al. 2002c).

As shown above, VIGS has the advantages of simple operation, high silencing efficiency, and labor saving when compared with transgenic methods. Hence, VIGS has been widely used in plant gene function research. To date, the VIGS system has been established in a variety of plants, including tomato (Holzberg et al. 2002), strawberry (Jia et al. 2011), peach (Turnage et al. 2002), barley (Holzberg et al. 2002), pepper (Chung et al. 2004), legume species (Constantin et al. 2004), potato (Faivre-Rampant et al. 2004), and *Petunia* (Constantin et al. 2004). Some researchers have begun to study fruits. Fruit is the reproductive organ of certain plants and has important economic value. Research applying VIGS in fruit will help clarify the fruit-ripening mechanism.

2.2.1 The Basic Principle of VIGS

VIGS is a posttranscriptional gene silencing (PTGS)-based technique that exploits the defense mechanisms employed by plants to prevent intracellular viral proliferation and immobilize extracellular viruses (Gray et al. 1992). After the *Agrobacterium tumefaciens*-carrying virus genome and part of the plant target gene infect a plant, the modified RNA is transcribed in plant cells and then replicated by endogenous RNA-dependent RNA polymerase (RdRp) to produce double-stranded RNAs (dsRNAs) and trigger PTGS. The main form of plant PTGS (Nishikura 2001) is that the plant uses its posttranscriptional RNA silencing mechanism to recognize and degrade the target gene dsRNA and produce a large number of siRNAs (Bernstein et al. 2001). Dicer is a cleaving enzyme that recognizes dsRNAs from their ends and cleaves them into siRNAs approximately 21–23 nucleotides in length (Zamore et al. 2000). As guiding molecules in plants, siRNAs form RNA-induced silencing complex (RISC), a protein complex approximately 500 kDa in size (Hammond et al. 2000), through interactions between RNA and protein (Burch Smith et al. 2004). Studies have shown that RISC has an exonuclease activity and can recognize homologous RNA to silence the RNA of the target gene (Robertson 2004). In addition, RISC acts as a helicase to assist in unwinding dsRNAs and activating siRNA (Baulcombe 2004). A single-stranded siRNA recognizes and degrades the target gene mRNA with a complementary sequence in plants (Nykänen et al. 2001). To enhance RNA silencing, siRNAs act as an amplification primer to synthesize dsRNA with single-stranded RNA as a template under the guidance of RDRP (Burch Smith et al. 2004). The resynthesis of dsRNA provides substrates for the production of more siRNA, thereby realizing the amplification of the RNA silencing signal and enhancing RNA silencing reaction. This process drives the continuation of signal amplification reaction in plants (Benedito et al. 2004). In addition, when a seedling of the control group (wild type) was grafted onto a plant of the RNA-silenced group, RNA silencing of the target gene also appeared in the control plant as the plant grew, suggesting the mobility of RNA silencing signals in plants (Benedito et al. 2004).

2.2.2 The Methods for VIGS

2.2.2.1 Types of VIGS Viral Vectors

Usually, it takes only a few weeks to construct VIGS viral vectors and observe the phenotype of infected plants; therefore, VIGS can serve as a bridge, easily and quickly linking a gene that has a known sequence but unknown function to its biological functions. The characteristics of a viral vector determine the area affected by VIGS in plants. Viral vectors are transported from the infected area to most of other tissues through the transportation system in the plant, but most viruses cannot infect the growth points in plants (Peele et al. 2001). The ability of a viral vector to induce gene silencing in plants is determined by multiple characteristic factors, including the ability to infect growth points and induce gene silencing in plants, the size of the virus genome, nucleic acid types and genome composition of the viral vector, and the host environment. In addition, use of viral vectors must take into account the safety risk to operators and the environment. Several plant viruses have been used for VIGS, including some positive-strand RNA viruses, such as potato virus X (PVX), tobacco mosaic virus (TMV), and TRV (Purkayastha and Dasgupta 2009). TMV is the first virus to be used as a VIGS viral vector in plants (Kumagai et al. 1995). TRV can effectively carry silencing signals to the growth points in plants, providing the possibility to study gene function in plant developmental organs (Ratcliff et al. 2001). TRV is a double-stranded RNA virus consisting of two single-stranded RNAs: RNA1 and RNA2. pTRV-RNA1 contains genes related to virus replication and mobility. RNA2 encodes the coat protein and serves as a site for insertion of target gene fragments (Angenent et al. 1986; MacFarlane 1999). Taking advantage of the strong infective ability of *Agrobacterium tumefaciens*, the use of bivalent expression vectors carrying the TRV genome greatly eases the viral infection procedure for VIGS (Liu et al. 2002c). In *Agrobacterium*-mediated genetic transformation, two TRV genome structures are transferred into *Agrobacteria* separately with the target gene fragments cloned into RNA2, and then the mixture of the two groups of *Agrobacteria* is used to coinfect the leaves of the plant. Compared with other viral vectors, the TRV vector possesses several advantages. For example, it can easily and quickly infect plant leaves and generate the desired gene-silencing phenotype within a few days after infection, it can infect the growth points of a plant and induce gene silencing in reproductive organs, and it can be combined with the Gateway cloning technology to achieve high-throughput functional analysis of target genes (Liu et al. 2002c). To date, application of the TRV vector has achieved good results in tobacco and Solanaceae plants (e.g., tomato), but it is not suitable for model species such as *Arabidopsis thaliana*. In addition, RNA-virus-derived VIGS systems with satellite and helper RNAs demonstrate a higher silencing efficiency than those consisting of only satellite RNA viruses. Replication and movement of satellite viruses in plants highly rely on the genome information of helper viruses. A recently reported VIGS system based on satellite tobacco mosaic virus (STMV) uses the U2 strain of TMV (TMV-U2) as a helper virus (Gosselé et al. 2002).

Another source of VIGS vectors is bidirectional single-stranded DNA viruses and single-component DNA viruses that require helper or satellite DNA. For example, tomato yellow leaf curl China virus (TYLCCV) and its associated DNA β satellites can introduce target genes and cause gene silencing in plants (Tao and Zhou 2004). However, DNA viruses cannot be widely used as gene-silencing vectors because their large genome structure may limit the mobility of viruses (Palmer and Rybicki 2001). The geminivirus vector can infect plant tissues and is a very promising viral vector (Fofana et al. 2004). The bipartite geminivirus cabbage leaf curl virus (CbLCV) vector, which is the first viral vector used in *Arabidopsis thaliana*, also has high potential. The tomato golden mosaic virus (TGMV) vector can effectively induce silencing of proliferating cell nuclear antigen (*PCNA*) gene (Peele et al. 2001). *PCNA* is a key gene required for DNA replication in the growth points in tobacco, and its transgenic plants cannot survive (Kjemtrup et al. 1998). TGMV and cabbage leaf curl virus (CaLCuV) induce RNA silencing of target genes through DNA replication mechanisms in nuclei of host plants. Recently, an easier-to-perform and more effective VIGS system was established by using rice tungro bacillus virus (RTBV), a DNA virus from infected rice, which achieved a silencing rate of 90% for phytoene desaturase (*PDS*) gene in rice leaves. *PDS* is generally selected as a reporter gene to assess the VIGS performance in different plants because the *PDS* gene-silencing phenotype can be easily observed by the naked eye.

2.2.2.2 Infection Methods for VIGS

In addition to a good viral vector, a suitable infection method is key to achieving satisfactory gene silencing outcomes. At present, there are three main infection methods proposed in the existing literature (Liu et al. 2021). The first is through mechanical wounding. The virus RNA vector transcribed *in vitro* or extracted from the infected leaves infects the leaf wound created by using a sterile toothpick or quartz sand, e.g., early-stage PVX infection. Although the operation is complicated, this method often reaches a satisfactory gene silencing outcome, making it suitable for plants that are difficult to infect by other methods, such as *Arabidopsis thaliana* (Ratcliff et al. 2001). The second infection method is to use *Agrobacterium*, which is highly infective in dicotyledons. Along with the infection of *Agrobacterium*, the viral vector carrying the target gene enters and establishes a transient expression system for successful gene silencing in the plant tissue. *Agrobacterium* infection of plant leaves is achieved mainly by injection with a syringe without a needle, high-pressure spraying on the leaf surface, or vacuum infiltration into leaf tissue (Figure 2.1) (Burch Smith et al. 2004). This method can be applied to infect plants with either DNA or RNA viruses and can successfully establish a transient expression system for exogenous proteins in plants. The

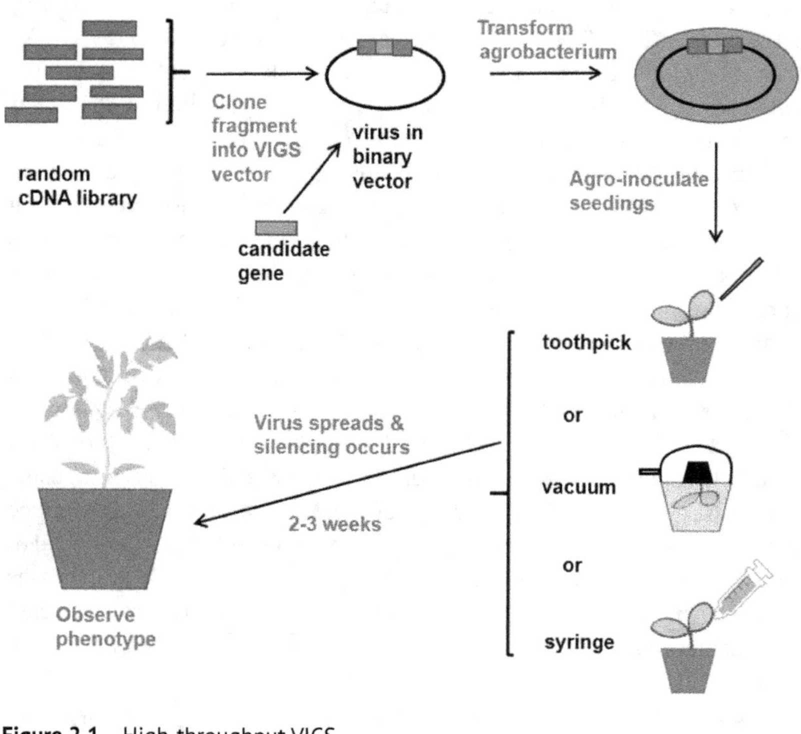

Figure 2.1 High-throughput VIGS.

third infection method is particle bombardment, which is also a good approach to introduce viral vectors into plants and is mainly used for DNA viruses (Ratcliff et al. 2001). The infection method should be properly selected based on the characteristics of plant materials and viral vectors.

2.2.3 Application of VIGS

VIGS technology is a very sophisticated method for gene function identification and has been used to identify a variety of genes in plants (Chanderbali et al. 2009; Muruganantham et al. 2009; Chai et al. 2011; Pan et al. 2013b; Li et al. 2022; Rössner et al. 2022). This section describes the relevant applications of VIGS vector systems in various plants and provides reference information to researchers in the related field.

There are only a few VIGS systems applicable for monocotyledonous plants (Table 2.1). Barley stripe mosaic viruses (BSMVs) have been successfully used to induce gene silencing in monocots, including barley, wheat, rice, maize, and some other cereal crops (Holzberg et al. 2002). In addition, BSMV can be used to silence the genes in both aboveground tissues and underground root tissue, providing an effective tool for research on grass plants. However, not all root genes can be silenced because they vary in stability after being inserted into viruses. In addition to Poales species, other monocots (e.g., ginger, a tropical crop) are also sensitive to the BSMV-VIGS system; hence, this system is very promising for the research of tropical plants, such as banana and plantain.

VIGS systems have been widely used in dicotyledonous plants and have laid a good methodological foundation for the research of fruit development and ripening mechanism (Table 2.2). Apple latent spherical virus (ALSV) is a unique virus that can be used to establish an ALSV-VIGS system effectively silencing endogenous

Table 2.1 Monocotyledonous plants suitable for VIGS.

	Virus-derived vector	Reference article
Brachypodium distachyon	BSMV	(Pacak et al. 2010)
Costus spicatus	BSMV	(Renner et al. 2009)
Hordeum vulgare	BSMV, BMV	(Holzberg et al. 2002)
Oryza sativa BMV	BMV, RTBV	(Ding et al. 2006)
Phalaenopsis spec	BMV, RTBV	(Lu et al. 2007)
Triticum aestivum	BSMV	(Scofield et al. 2005)
Zea mays	BMV	(Ding et al. 2006)
Hordeum vulgare L.	FoMV	(Liu et al. 2016)
Zea mays	FoMV	(Mei et al. 2016)

Table 2.2 Dicotyledonous plants suitable for VIGS.

	Virus-derived vector	Reference article
Fragaria ananassa	TRV	(Chai et al. 2011)
Malus domestica	ALSV	(Chai et al. 2011)
Pyrus ssp.	ALSV	(Chai et al. 2011)
Cucumis sativus	ALSV	(Igarashi et al. 2009)
Citrullus lanatus	ALSV	(Igarashi et al. 2009)
Lagenaria siceraria	ALSV	(Igarashi et al. 2009)
Capsicum annum	TRV	(Chung et al. 2004)
Solanum lycopersicon	TRV ASLV	(Liu et al. 2002c; Brigneti et al. 2004)
Vaccinium myrtillus	TRV	(Jaakola et al. 2010)
Glycine max	BPMV	(Kikuchi et al. 2016)
Lycopersicon esculentum Mill	TRV	(Fantini and Giuliano 2016)
Capsicum annuum	TRV	(Kim et al. 2017)
Subg. Persica L	PNRSV	(Cui and Wang 2017)

genes in many plants, such as *Pisum sativum* and *Malus Domestica*. Sasaki et al. introduced the ALSV-VIGS vector into the seedlings of *Malus domestica*, *Pisum sativum*, and Japanese *Pisum sativum* via particle bombardment and generated prominent gene-silenced phenotypes (Sasaki et al. 2011). The VIGS system was first developed in *Nicotiana benthamiana*. It has been reported that the vector carrying the hybrid gene sequence of TMV and tomato mosaic virus (ToMV) can more effectively silence the *PDS* gene in *Nicotiana benthamiana* (Kumagai et al. 1995). Later, other vectors were developed for VIGS. The TRV vector is most commonly used for studying Solanaceae plants and has been widely used in tomato, potato, eggplant, and pepper. Recent research has shown that directly injecting *Agrobacterium tumefaciens* infection medium into citrus fruit can silence target genes to investigate the functional genes related to anthocyanin accumulation.

VIGS is widely used. In 2005, a VIGS vector derived from Poplar mosaic virus (PopMV) was developed and successfully used to silence transgenes in *Nicotiana* species (Naylor et al. 2005). However, whether this VIGS system can be used in poplar has not been established yet. A VIGS system using the geminivirus beet curly top virus (BCTV) has been established in spinach (Sather et al. 2010). The ALSV-VIGS system is very effective for silencing genes in Fabaceae plants, including *Glycine max* (soybean), *Pisum sativum* (pea), and two *Vigna* species (*V. unguiculata* and *V. angularis*). In cotton plants, the DNA geminivirus cotton leaf crumble virus (CLCrV) is used to construct a VIGS vector that can silence the genes in the integument tissue of cotton ovules to affect the synthesis of cotton fiber.

Most existing studies on tomato fruit ripening and development used VIGS as a study method (Ma et al. 2022).

2.3 Transgenic Technology

Transgenic technology introduces artificially isolated and modified nuclear genes into the genome and expresses these exogenous genes in an organism to genetically modify the heritable traits of the organism. This technology can be divided into artificial transgenic technology and natural transgenic technology according to the approach or divided into plant transgenic technology, animal transgenic technology, and microbial gene recombination technology in terms of the research object. Genes can be transferred by biological, chemical, and physical methods. Biological methods include *Agrobacterium*-mediated genetic transformation, the virus-mediated approach, and the pollen-tube pathway method. Chemical methods include polyethylene glycol (PEG)-mediated and liposome-mediated approaches. Physical methods include gene gun bombardment, microinjection, and electroporation.

2.3.1 Plant Transgenic Technology

2.3.1.1 Concept

Plant transgenic technology is a tool for transferring exogenous genes originating from any organism (e.g., animals, plants, or microorganisms) or artificially synthesized into the genome of the recipient plant by biological, physical, or chemical means to create traits that can be stably expressed and inherited against a new background.

2.3.1.2 Methods

Genes can be transferred by biological, chemical, and physical methods. Biological methods include *Agrobacterium*-mediated genetic transformation, the virus-mediated approach, and the pollen-tube pathway method. Chemical methods include PEG- and liposome-mediated approaches. Physical methods include gene gun bombardment, microinjection, and electroporation. The most commonly used methods are *Agrobacterium*-mediated genetic transformation and gene gun bombardment.

1) *Agrobacterium*-mediated genetic transformation

Agrobacterium is a gram-negative bacterium that is ubiquitous in soil. *Agrobacterium tumefaciens* and *Agrobacterium rhizogenes* are most commonly used for plant transgenic technology, and their cells contain Ti and Ri plasmids, respectively. The plasmids contain a DNA segment that can cause the occurrence of plant tumors after it is replicated, cleaved, transferred, and integrated into the plant genome and expressed in the plant through a specific mechanism, during which this DNA segment acts as T-DNA. After the engineered *Agrobacterium tumefaciens* infects a wound and enter the cells in the plant, T-DNA is inserted into the plant genome. *Agrobacterium* is a natural plant genetic transformation system and has been most commonly used for genetic transformation of both monocots and dicots. In recent years, it has been widely used in plants such as rice, maize, wheat, tomato, and *Arabidopsis thaliana*. Compared with other methods, *Agrobacterium*-mediated genetic transformation has many advantages, including simple operation, no need for special equipment, and a shorter culture period. It is a mature technology with a high transformation efficiency and stability due to the direct single-copy integration of most exogenous genes. This technology can control the specific expression of a target gene in certain organs by using different promoters. In addition, this technology can transfer large DNA fragments into the plant genome because it rarely causes gene silencing. However, because T-DNA can be inserted into any sites of the plant chromosome, it is possible that beneficial genes are inactivated.

2) Virus-mediated genetic transformation

The plant virus-mediated transgenic technique constructs a viral vector as carrier by inserting the exogenous target genes into the virus genome to introduce exogenous genes into plant cells and integrate them into the plant genome through viral infection. The three major plant viral vector systems are the single-stranded RNA virus, single-stranded DNA virus, and double-stranded DNA virus vector systems. However, it is worth noting that if the plant is also infected with other viruses that may exchange information with the vector virus, the virus used for vector construction might regain its ability to cause disease and become potentially harmful to the plant.

3) Pollen-tube pathway approach

The pollen-tube pathway transformation approach, also known as the ovary injection method, involves an injection of exogenous DNA into the placenta along the central axis of the ovary. As the fertilized eggs or embryo cells develop, the exogenous DNA is integrated into the genome of the recipient cell, causing the fertilized egg to develop into a new transgene-carrying individual plant. This technique realizes the genetic transformation at the plant level and can be directly conducted on plants during cultivation without nuclear preparation and tissue culture requirements. In addition, it can directly obtain the transformed seeds. The pollen-tube transformation approach is a new strategy for gene transformation in living whole plants. Its greatest advantage is that it does not rely on tissue culture to artificially regenerate plants, so it is easy to master and perform and does not require a well-equipped laboratory.

4) Gene gun approach

The gene gun was first invented by the research team led by Sanford from Cornell University in 1987, through which nucleic acid molecules attached to the surface of metal particles moving at a high speed are introduced into recipient cells for genetic transformation. This method has been applied in many plant species, including wheat, rice, corn, soybean, cotton, and tomato. The gene gun approach is essentially a physical process; therefore, its application is not limited by the genotypes of recipient plant materials. The time needed to obtain transgenic plants is shortened, and the transgenic plants obtained have fewer variations and usually have normal fertility. However, the exogenous DNA fragments injected by gene gun are hardly integrated into the chromosome of the recipient cell because the fragments can be easily recognized by the restriction-modification system in recipient cells. Therefore, the gene gun approach has a very low efficiency in introducing exogenous genes that can be stably inherited and expressed in plants.

5) Microinjection method for genetic transformation

The microinjection method for genetic transformation is a technique of microscopically injecting genetic material into cells with a microinjector and obtaining the transformed plants through tissue culture. The technique was first applied to animals; and in the mid-1980s, it was applied to plants. Although DNA injection is accurate and predictable and can overcome the difficulties encountered in distant hybridization, application of the microinjection method is limited by its low efficiency and unstable gene expression, and therefore, it has been gradually abandoned with development and application of new technologies.

6) Electroporation-mediated genetic transformation

The electroporation-mediated transformation technique was developed in the early 1980s. It uses high-voltage electric pulses to create transient reversible pores in the cell membrane, through which exogenous molecules are delivered into cells for genetic transformation. Laser-mediated transformation is a similar technique. It uses a high-energy laser beam at a defined wavelength focused within a diameter of approximately 0.3–1.5 μm to generate transient pores in the cell membrane. Through these pores, which automatically recover in a short time, exogenous DNA is introduced into cells. Both electroporation- and laser-mediated transformation techniques can be used for monocots and dicots. Although the two techniques can be used to genetically transform a wide range of plant materials, their application is substantially hampered due to their low transformation efficiency and numerous other issues.

7) PEG-induced genetic transformation

PEG can tightly bind to DNA to form PEG/DNA complexes through the interaction between charges, thereby enhancing the permeability of the protoplast membrane and easing the absorption of the PEG-DNA complexes. Due to its advantages of low cost, stable performance, good repeatability, and ability to be performed without special instruments, PEG-induced genetic transformation was widely used in the early days after its invention. However, this method has obvious drawbacks, such as considerable difficulties in protoplast culture and regeneration, genotype limitation, high variation of transformed plants derived from protoplast regeneration, high rate of albino seedlings, and low transformation efficiency.

8) Liposome-mediated genetic transformation

The liposome-mediated genetic transformation technique fuses plant protoplasts with negatively charged liposomes containing exogenous genes for gene transfer. This technique generally includes four steps: separation and purification of protoplasts; preparation of liposomes; fusion of liposomes and protoplasts; and culture, screening, and identification of transformants. At present, the liposome-mediated transformation technique is the most effective for animal cell transformation. However, because the common recipient material for liposome-mediated transformation is protoplasts, applications of this technique are very limited in plants whose protoplast culture is difficult. Therefore, this technique requires further improvement.

2.3.2 Application of Transgenic Technology

Transgenic technology is widely used in medicine, industry, agriculture, environmental protection, energy, new materials, and other fields. In recent years, with the development of molecular biology, the research on genetic regulation has been increasingly deepened, especially in the aspects of signal transduction, transcription factors, and RNA cleavage. Great progress has been made in the research on plant transcription factors. As the structure and function of transcription factors are further understood, it will be possible to artificially control the specific expression of plant genes to comprehensively improve plant traits in near future.

2.3.2.1 Transcription Factors and Transgenic Technology

Transcription factors, also known as trans-acting elements, are a group of factors that can specifically bind the cis-acting element in the promoter region of eukaryotic genes to ensure that the protein encoded by the target gene is expressed at a specific level in a specific time and space. When environmental conditions change, the signal that is perceived and transmitted in the plant results in the activation of certain transcription factors. Subsequently, the binding of the transcription factors with the corresponding cis-acting elements leads to the activation of the RNA polymerase II transcription complex and initiates the transcription and expression of specific genes that regulate the response of the plant to internal and external signals. The expression of many genes in plants is regulated by the interaction of specific transcription factors and cis-acting elements. During plant growth and development, the response of a plant to internal and external environmental changes involves the precise regulation of genes. For example, drought, high salt, hormonal stimuli, and diseases, as well as in vivo cellular developmental signals, can trigger the expression of corresponding transcription factors, i.e., trans-acting factors, which bind the cis-acting elements to activate the RNA polymerase II transcription complex and initiate gene transcription and expression. This process regulates and ensures the proper physiological and biochemical responses of the plant to external signals. Therefore, transcription factors play a key role in the regulation of gene expression. Knowledge of the structure of transcription factors and the detailed regulatory mechanism of transcription factors on the expression of genes, especially certain key genes, can lay a solid foundation for improving crop traits and achieving satisfactory genetic transformation performance in plants. The *Agrobacterium*-mediated genetic transformation technique, which has been widely used in research, is the most convenient and effective tool to study the biological functions of transcription factors in plants. Transgenic technology has been used to determine the biological functions of target transcription factors in a large number of studies.

2.3.2.2 Application of Transgenic Technology in Climacteric Fruits

Climacteric fruit is a very important type of fruit from higher plants. Growth and postharvest ripening of climacteric fruits are complex biological processes with transcription factors acting as key regulators. To date, the major biological functions and regulatory mechanisms of respiration in climacteric fruits have been revealed.

Tomato is a model plant for the research on climacteric fruit. Ripening of tomato fruit is a process involving the interaction and mutual regulation of various biological regulatory pathways, which ultimately leads to fruit ripening. Transcription factors are an indispensable regulator of tomato fruit ripening and play a key role during the ripening process.

In wild-type tomato fruits, the *RIN* gene is a key regulator of fruit ripening that encodes the MADS-RIN transcription factor and regulates the expression of hundreds of ripening-related genes. In wild-type AC tomato transgenic plants that overexpress *RIN-MC* gene, fruit ripening is severely inhibited. Downregulation of *RIN-MC* gene expression in *rin* mutants causes the fruit to turn yellow and show a light red color. In a study comparing *rin* mutant and *rin 35S::RIN-MC* RNAi transgenic fruits, transcriptomic analysis revealed the presence of many relevant genes, including *RIN-MC*. The researchers also found that the fused *RIN-MC* gene can be translated into protein in the nucleus, which can act as a transcription factor and interact with other MADS-box transcription factors. This study revealed the important role of *RIN-MC* in forming the fruit phenotype of tomato *rin* mutants and comprehensively investigated the mechanism causing nonripening fruit in *rin* mutants (Li et al. 2018).

In another study, the researchers used the experimental transgenic technology and revealed that the transcription factor *RIN* regulated the metabolism of sucrose and the ripening process of tomato fruit. In addition, the zinc-finger transcription factor *SlZFP2* can negatively regulate abscisic acid (ABA) biosynthesis and affect tomato fruit ripening (Weng et al. 2015).

The *TAGL1* gene of the MADS family is a key regulator of tomato reproductive development and is mainly expressed at the stages of flowering and early fruit growth and development. Silencing *TAGL1* by RNAi may lead to remarkable cuticle changes, including the reduced thickness, hardness, and content of the cuticle. Transcription factor *TAGL1* regulates cuticle growth and development through regulating the biosynthetic activity of epidermal cells. Formation of the cuticle is a complex and coordinated process during expansion and ripening of fleshy fruit such as tomato.

Tomato fruit growth and development are affected by endogenous hormones, gene regulatory factors, and external factors. The NAC family is a very important transcription factor family. The transcription factor *SlNAC4* is highly expressed in the calyx at the initial fruit-ripening stage, and its expression can be increased by various stresses. Suppressing the expression of *SlNAC4* by RNAi led to a delay of tomato fruit ripening, and other ripening-related genes in the transgenic tomato obtained also had a lower expression level. Therefore, the transcription factor SlNAC4 is a positive regulator for tomato fruit ripening and may act downstream of *RIN* regulation (Zhu et al. 2014).

Fruit growth, development, and ripening require critical biological regulatory mechanisms, which are key to plant growth and development. Nguyen et al. found that the transcription factor *GOLDEN2-LIKE* (*GLK*) plays a regulatory role in tomato fruit growth and ripening (Nguyen et al. 2014). The number of chromoplasts and the content of chlorophyll are influenced by numerous environmental and genetic factors, and they are positively correlated with the levels of photosynthesis and photosynthetic products. *GLK* can regulate the levels of chromoplasts and chlorophyll. Tomato *GLK* includes *GLK1* and *GLK2*, which are distributed in leaves and fruits, respectively. Mutant and transgenic analysis indicated that the peptides encoded by these two genes had similar functions.

In addition, stomatal closure is part of the innate immune response of plants to resist the invasion of pathogens. However, this defense mechanism can be hampered by coronatine (COR), which simulates the effect of jasmonic acid (JA) to reopen the stomata, allowing the entry of bacteria. ABA plays an important role in stomatal opening and closing. In a study exploring the relationship of two NAC family transcription factors, *JA2* and *JA2L*, with stomatal behavior in tomato by using transgenic technology, ABS activated the expression of *JA2*, which acted as a positive regulator of ABA-controlled stomatal closure. In other words, the regulatory role of *JA2* is implemented through regulating the expression of genes related to the ABA biosynthesis pathway. In contrast, JA and COR can activate the expression of *JA2L*, which plays a regulatory role by regulating salicylic acid metabolism (Du et al. 2014).

2.3.2.3 Application of Transgenic Technology in Nonclimacteric Fruits

With the development and application of transgenic technology, a large number of transcription factors related to fruit ripening and plant growth were also found in banana, another type of climacteric fruit.

MaERF11, a transcription factor in the ethylene synthesis pathway, negatively regulates the ripening of banana fruit. *MaERF11* binds to the histone deacetylase MaHDA1 and reduces the expression of its target gene *MaACO1* and ripening-related protein, thereby affecting the ripening of banana fruit.

Transcription factors MusaVND2 and MusaVND3 are localized in the nucleus and encoded by the vascular-related NAC domain genes. In the embryonic cells of banana, lignification gradually induces the increase in the transcription levels of *MusaVND2* and *MusaVND3*. Studies have found that banana embryonic cells differentiated into abnormal tubular cells after overexpression of *MusaVND2* and *MusaVND3*. This abnormal cell differentiation occurred in the cortical cells of bulbs and the epidermal and mesophyll cells of leaves in the transgenic banana. In addition, large amounts of lignin and fibrinogen were detected in the transgenic banana. All the experimental results above demonstrated that MusaVND2 and MusaVND3 regulate secondary wall deposition (Negi et al. 2016).

Similarly, the transcription factors of the MaMADS family have important functions in banana fruit. These transcription factors can inhibit or delay fruit ripening, thereby prolonging the shelf life and improving fruit quality. The underlying mechanism is similar to that in tomato (Elitzur et al. 2016).

In grape plants, low-temperature treatment can rapidly change the expression levels of temperature-sensitive transcription factors *VvCBF2*, *VvCBF4*, *VvCBFL*, and *VvZFPL* in leaves, flowers, and stems. Some studies have found that *Arabidopsis thaliana* overexpressing *VvCBF2* and *VvZFPL* had smaller rosette leaves and lower chlorophyll content, which resulted in a lighter green color of leaves and eventually slower growth than that of normal plants. This finding indicates that *VvCBF2* and *VvZFPL* regulate the synthesis and growth of chlorophyll in grape plants. In addition, transgenic plants overexpressing *VvCBF2*, *VvCBF4*, *VvCBFL*, and *VvZFPL* had a higher cold tolerance, suggesting that these genes can enhance cold tolerance in grape plants (Takuhara et al. 2011).

2.3.2.4 Application of Transgenic Technology in Other Plants

A large number of studies have also used transgenic technology to study transcription factors in other plants, especially model plants such as *Arabidopsis*, maize, and rice, which are suitable for *Agrobacterium*-mediated genetic transformation.

Plants can self-adapt to different environmental stresses. Transcription factor NAC is involved in the response of plants to abiotic stress. *Arabidopsis* plants overexpressing *CiNAC3* and *CiNAC4* are less sensitive to ABA during seed germination and have a higher salt tolerance. *CiNAC4* promotes the formation of lateral roots and responds to salinity changes and ABA treatment (Han et al. 2015).

In a study comparing wild-type and transgenic *Petunia* plants, the expression levels of transcription factors *B-Peru+Mpap1* and *RsMYB1*, which are anthocyanin regulators, were significantly higher in transgenic *Petunia* than in wild-type plants. The results indicate that sucrose-induced anthocyanin accumulation in the growing tissues of *Petunia* requires the involvement of transcription factors regulating anthocyanin synthesis (Ai et al. 2016).

2.3.3 Development of New Technologies

In addition to the common transgenic method, including RNAi interference and overexpression, scientists have put considerable effort into developing new plasmids and new methods to expand the scope and field of research.

To further investigate transcription factors in different life stages of plants, an artificial small interfering peptide (a-siPEP) approach has been developed based on the overexpression of specific protein domains. This research has demonstrated that a-siPEPs attenuate nuclear import and DNA binding of target transcription factors. Hence, the a-siPEP tool is an efficient genetic means of weakening the role of specific transcription factors in plants (Seo et al. 2012).

2.4 Gene Editing

2.4.1 Concept

Gene editing refers to the technology of "modifying" a target gene to affect its function, which is an effective method to study the biological function of the gene. At present, the most commonly used "modifications" include gene knockout, mutation, and overexpression. In conventional forward genetics, chemical-induced mutation or transposon-mediated mutation is applied to study gene function. However, this strategy has an obvious limitation in its targeting of nonspecific modification sites. In contrast, homologous recombination (HR)-mediated gene knockout or knock-in can precisely modify the target sites. However, the practical application of this approach is limited by its onerous operation and low efficiency. This challenge prompted the emergence of genome-editing technology, an experimental tool that can efficiently edit specific gene sequences.

Genome-editing technology uses exogenous DNA to modify the genome at specific sites of the chromosome in cells and obtain an organism with the expected genetic changes in the genome sequence. This technology can make modifications precisely at the target sites in the genome by using specific nucleases. The double-stranded DNA is cleaved at the target site to produce double-stranded DNA breaks (DSBs), which are then repaired through nonhomologous end joining (NHEJ) or HR. Meanwhile, different mutation types can be generated by editing the sequence near the DNA breakpoint and using the plant's own repair mechanism. NHEJ directly connects the broken ends of the double-stranded DNA, and during the repair process, the errors that easily occur lead to various mutations, e.g., base insertion or deletion mutation or frameshift mutation that causes gene dysfunction, to achieve target gene knockout. After breaking the double-stranded DNA, a specific sequence segment can be inserted (knock-in) or deleted (knockout), or a specific base or sequence can be mutated at the breakpoint by introducing the designed donor DNA template and using an HR mechanism. This true "gene-editing" process modifies and edits the original genome while simulating the natural gene mutation processes. Compared with the traditional gene-targeting technology based on HR technology, the new gene-editing technology retains the characteristics of site-specific modification but provides a higher mutation efficiency with a shorter construction time and lower cost.

Three currently widely used gene-editing tools are zinc-finger nucleases (ZFNs), transcription activator-like effector nucleases (TALENs), and RNA-guided clustered, regularly interspaced, short palindromic repeats–CRISPR-associated proteins (CRISPR/Cas) (Figure 2.2).

2.4.2 Principles

2.4.2.1 ZFN Technology

A ZFN is composed of zinc finger proteins (ZFPs) and nuclease Fok I. The DNA-recognition domain composed of ZFPs can recognize and bind to specific sites,

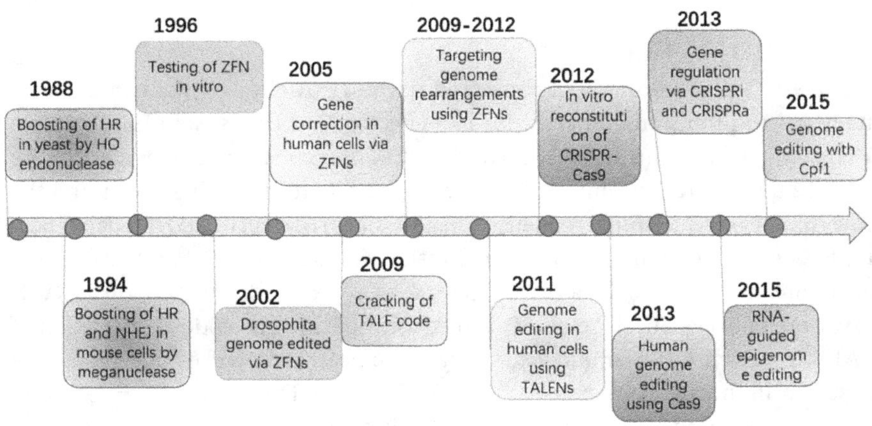

Figure 2.2 Timeline of gene-editing technology.

and the DNA-cleaving domain composed of Fok I can cleave DNA at the target site to generate DSBs (Xiao et al. 2011). DSBs in plant cells can be repaired through HR and NHEJ. HR may make either repair or insertion modifications at the target sites, while NHEJ repair is more insertion- or deletion-prone. Both can cause frameshift mutation to achieve the purpose of gene knockout. Fok I must dimerize in order to function (Bitinaite et al. 1998), which greatly reduces off-target risks caused by random cleavage. To construct ZFNs targeting specific sites, users only need to design 8–10 zinc finger domains for the target gene and then fuse them with nuclease (Klug 2010). After the obtained ZFNs are transported to the cellular nuclei in the wounded tissue through *Agrobacterium*-mediated genetic transformation (leaf disc infiltration approach), the two complementary ZFNs can recognize the DNA strand at the target site along both the 5′→3′ and 3′→5′ directions in a tail-to-tail manner through the zinc fingers. Subsequently, two Fok I nucleases form a dimer so that the enzyme activity is activated to cleave DNA at the target site and form DSBs, which triggers the DNA damage repair mechanism (Porteus and Carroll 2005; Kim et al. 2011).

The key of ZFN technology is to construct a ZFP domain that specifically recognizes and targets specific DNA sequences. Zinc finger is a protein motif that widely exists in eukaryotes. The ZFP domain consists of a series of Cys2/His2 ZFPs connected in tandem. Each ZFP contains approximately 30 amino acid residues, which are folded to form the secondary structure of α-β-β (Wolfe et al. 2000). The binding specificity of a ZFP to DNA is determined by the seven amino acid residues at position −1 to +6 on the α helix. Multiple ZFPs are connected in tandem to specifically recognize a DNA sequence. The residues at positions −1, +3, and +6 can directly interact with a base triplet of the target sequence; therefore, a change in any one of these three residues will affect the recognition of the target base by the ZFP. In addition, the ZFPs interact with each other. All effects described above are collectively referred to as context-dependent effects, which will affect the specific binding of ZFP to target sites.

2.4.2.2 TALEN Technology

TALEN is a gene-editing tool composed of the DNA-cleaving domain of Fok I nuclease and transcription activator-like effector (TALE) instead of ZFPs as the DNA-binding domain. TALE is derived from *Xanthomonas* and consists of more than 12 tandem repeat units that can recognize specific DNA base pairs and the N-terminal and C-terminal sequences on both sides (Boch et al. 2009). Each repeat unit generally contains 34 amino acid residues, where the 12th and 13th amino acids are highly variable and referred to as repeat-variable diresidue (RVD) (Moscou and Bogdanove 2009). RVD mediates specific nucleotide recognition of TALE in which a two-amino-acid repeat specifically recognizes and binds one base pair in the following pattern: NI = A; NG = T; HD = C; and NN = G. Based on this recognition pattern, a TALE protein that specifically recognizes the target

DNA sequence can be customized by connecting and cloning multiple TALE repeat units in series based on the length of the target DNA sequence of interest. Two TALE recognition modules are constructed based on two adjacent target sequences (16–20 bp) with a spacer of approximately 13–22 bases wherein the dimerizing Fok I domains introduce a DSB. Subsequently, the two recognition modules are fused to the N termini of Fok I to form a eukaryotic expression vector, namely, a TALEN plasmid pair. Subsequently, the TALEN plasmid pair is used to transform cells and coexpress the fused proteins that can bind to the target sites, allowing the dimerized Fok I to cleave the target sequence for gene editing (Li et al. 2011; Miller et al. 2011). Because the amino acid repeat module of TALE can specifically bind to a single base, theoretically, any target DNA sequence can be designed and recognized.

2.4.2.3 CRISPR-Cas System

CRISPR-Cas refers to clustered, regularly interspaced, short palindromic repeat sequences found in the DNA of some bacteria and archaea. These sequences have a length of approximately 24–47 bp and consist of direct repeats (DRs) and spacers. CRISPR-associated (Cas) nuclease (e.g., Cas9) is complexed with a segment of target DNA-specific single-guide RNA (sgRNA) (20 nucleotides in length) to form RNA-guided CRISPR-Cas nucleases that function as an immune system. The DR regions of CRISPR are mostly highly conserved within the same species; some spacer sequences are highly homologous to exogenous genomes such as phages and large plasmids (Bolotin et al. 2005; Mojica et al. 2005; Pourcel et al. 2005) and are named protospacers. sgRNA is formed by two noncoding RNAs, CRISPR RNA (crRNA) and trans-activating crRNA (tracrRNA), through base pairing (Jinek et al. 2012). Cas9 and sgRNA form a complex to cleave target double-stranded DNA sites defined by the proto-spacer adjacent motif (PAM) (the PAM sequence must be NGG, i.e., 5′-GN 19 NGG-3′ for Cas9). Cas9 contains an McrA-like HNH nuclease domain and an RuvC-like nuclease domain; therefore, the two domains cleave the complementary and noncomplementary (nontarget) strands, respectively, at the recognized target site. RuvC/RNase H-like fold nucleases include many proteins that can act on DNA and RNA (e.g., RNase H, RuvC, DNA transposase, retroviral integrase, and the Argonaut PIWI domain), showing a broad spectrum of nucleolytic functions (Makarova et al. 2011; Hsu et al. 2013). Through this adaptive immune defense system formed in the long evolutionary process, the CRISPR-containing bacteria or archaea can copy and insert a gene segment of the intruder (e.g., the invading virus, exogenous DNA, or phages) into their own DNA and pass along the change to their offspring. As a result, when encountering an invasion of the same type of intruders, endogenous Cas9 can cleave the intruder's DNA at the target site under the guidance of sgRNA to resist or eliminate the intruder. During gene editing of recombinant *Streptococcus pneumoniae* and *Escherichia coli* genomes, it was found that the cleavage would discontinue even when the target sequence had one different nucleotide out of 12 nucleotides near the PAM motif, but a mismatch of only five nucleotides would not affect the cleavage of an off-target DNA fragment (Fu et al. 2013). Coexpression of sgRNA and Cas9 without nuclease activity generates a DNA-recognition complex that can specifically interfere with transcription. This system, named CRISPR interference (CRISPRi), can effectively and reversibly inhibit the expression of single or multiple target genes (Qi et al. 2013). CRISPRi can be used to inhibit gene expression in mammalian cells.

The mechanism of the CRISPR-Cas system includes three stages: acquisition of the highly variable spacer region of CRISPR, expression of the CRIPSR locus (including transcriptional processing and posttranscriptional maturation), and cleavage or interfering with exogenous genetic substances of the CRISPR-Cas system. In the first stage, a small DNA segment copied from the invading phage or a DNA plasmid is integrated into the genome of the host bacterium to obtain the highly variable spacer of CRISPR. The DNA sequence of interest is inserted between two repeats at the 5′ end of CRISPR. Therefore, the arrangement of the spacer sequence in the CRISPR locus from 5′ to 3′ reflects the time sequence of exogenous genetic material invasions. The DNA sequence of the phage or plasmid corresponding to the spacer sequence is called a protospacer. Generally, there is a PAM region, which consists of several highly conserved bases, extending from the 5′ or 3′ end of the protospacer. PAM is 2–5 bases in length and 1–4 bases away from the protospacer (Mojica et al. 2009). New spacer acquisition can be divided into three steps. The first step is to identify the exogenous nucleic acids and determine the potential PAM of the exogenous DNA to select the sequences adjacent to PAM as candidate protospacers. The second step is to synthesize the repeats at the 5′ end of the CRISPR locus. The last step is to integrate the new spacer sequence between the two repeats. At present, only the first step has been confirmed. Cas1 and Cas2 proteins are the core proteins responsible for new spacer acquisition (Barrangou et al. 2007; Beloglazova et al. 2008). In addition, Csn2 protein in the Type II CRISPR/Cas system is also necessary for new spacer acquisition (Barrangou et al. 2007; Brouns et al. 2008).

Genome editing in eukaryotes can be accomplished by introducing the ectopic expression of two CRISPR/Cas9 components, guide RNA (gRNA) and Cas9 nuclease, into the same cell. A naturally occurring crRNA complementary to the target DNA sequence and a tracrRNA form a chimeric gRNA, which is a structural bridge between crRNA and Cas9. Cas9 is an RNA-guided DNA nuclease containing two twin catalytically active nuclease domains (NHN and RuvC). Cas9 is transported to a DNA sequence by complexing with a scaffold structure in the gRNA. The gRNA spacer region can then anneal to a complementary DNA sequence with Cas9 and create a DSB, as long as the target sequence neighbors a PAM. Cas9 orthologs may differ in their requirements for gRNA scaffold structures and PAM sequences in order to efficiently generate a DSB at the target site.

The most commonly used Cas9 is derived from *Streptococcus pyogenes* and requires an NGG (N, any nucleotide; G, guanine) PAM sequence for DNA targeting. These DSBs are repaired by evolutionarily conserved DNA repair pathways. The main pathway for DSB repair is NHEJ, which is error-prone and leads to the introduction of insertion or deletion (indel) at the target site . These mutations can disrupt genes by creating a premature stop codon in the open reading frame, leading to nonsense-mediated decay of the transcript. Alternatively, when a DNA template homologous to the target sequence is available, new sequences can be introduced by the homology-directed repair (HDR) pathway.

2.4.3 Methods

2.4.3.1 Construction of ZFN Expression Vectors

Currently, there are two main methods to construct a ZF domain. One is the "module assembly method" (Wright et al. 2006), which takes a ZF that can identify and bind a single base triplet as a module and then connects and assembles different modules successively according to the target sequence. This method is relatively simple but does not consider context-dependent effects. Hence, it has a high risk of off-target effects, which may lead to high cytotoxicity and overall low assembly efficiency. The other is the open "oligomerized pool engineering" (OPEN) method (Maeder et al. 2008). Among the shared ZF libraries, each ZF library corresponds to only one specific base triplet but consists of many ZF domains constructed with different amino acid sequences. Combining different ZF libraries will generate hundreds of thousands of ZF combinations, among which those with high affinity can activate the genes for downstream drug screening and gain a competitive advantage over others when cultured on selection medium. Finally, the highly efficient zinc finger domain can be identified through bacterial two-hybrid (B2H) screening. This method was later replaced by software-driven context-dependent assembly (CoDA), which avoids cumbersome protein library screening but has a lower specificity than OPEN.

2.4.3.2 Construction of TALEN Expression Vectors

The process of constructing TALE involves the following main steps: TALEN site design, TALE assembly, TALEN construction, linearization, ultrathin purification, in vitro transcription and recovery of RNA, concentration determination, microinjection, and detection of results. The choice of a single restriction enzyme cleavage site within the target DNA fragment is the key to success.

1) The full-length sequence synthesis method

The full-length sequence synthesis method directly synthesizes the coding sequence corresponding to the TALE repeat sequence according to the target gene sequence to be modified and then clones the sequence into the Fok I expression vector. This method is the simplest and fastest way to construct TALEN expression vectors, but its cost is also the highest. In 2010, Sangamo Therapeutics, Inc., first reported that the TALEN expression vector constructed by this method can effectively cleave DNA. The research showed that this method generates discrete edits or small deletions within endogenous human NTF3 and CCR5 genes at efficiencies of up to 25% (Miller et al. 2011). Subsequently, Mussolino et al. further optimized the parameters of TALEN expression vectors for effective DNA cleavage and increased the editing efficiency to 45% for endogenous human CCR5 and IL2RG genes. In addition, Mussolino et al. proposed that compared with ZFN, TALEN had similar gene-editing activity but significantly reduced nuclease-related cytotoxicity and off-target effects (Mussolino et al. 2011). However, each TALE repeat unit contains approximately 33–35 amino acid residues, requiring its coding DNA to be approximately 100 bp in length. Meanwhile, to ensure a high binding specificity, TALEN typically requires the target sequence to be at least 10–20 bp in length. As a result, the length of the sequence to be synthesized is in the range of 1–2 kb. Under the current technical conditions, it is possible to directly synthesize a DNA fragment with such a length, but the cost is too high, which has hindered the broad application of this method.

2) The modular assembly method

The modular assembly method, also known as the Golden Gate method, is a cloning method mediated by type IIS restriction endonuclease. Li et al. exploited four prevalent TAL repeats and their DNA recognition cipher to develop a "modular assembly" method for the production of TALENs (dTALENs) that are easier to use and can specifically recognize 23 consecutive bases in any gene, thereby greatly reducing the possibility of off-target effects (Li et al. 2011). This method is currently one of the most widely used methods for TALEN vector construction. Because the sites for recognition and cleavage are different, type IIS restriction endonuclease can specifically recognize the recognition site and nonspecifically cleave its downstream to produce a sticky end at the 5′ end of the target double-stranded DNA. Based on this characteristic, the base sequence at the sticky end can be artificially designed to cleave off the recognition site so that the recognition site will not be introduced into the cloned sequence. Hence, a series of vectors or polymerase chain reaction (PCR) primers with type IIS restriction endonuclease recognition sites can be designed to ensure that the sticky ends of TALE repeat units formed by IIS cleavage are different from each other but can pair according to the designed order. Through pairing of these sticky ends, the DNA sequences encoding different TALE repeat units can be spliced in order (Cermak et al. 2011). According to the origin of TALE repeat units for splicing, the Golden Gate method includes PCR-based (GG-PCR) and vector-based (GG-Vector) methods. Taking advantage of the degeneracy of the genetic codes for two adjacent amino acid residues in the repeat unit, the two methods use primers to design different sticky ends and introduce the same enzyme recognition site; subsequently, the PCR product or the DNA sequence encoding the TALE unit cloned into the vector is directly cleaved to obtain the coding sequence of the TALE unit needed for subsequent splicing.

Zhang et al. synthesized 17 TALEs that were customized to recognize specific DNA-binding sites and demonstrated that they can specifically modulate the transcription of endogenous genes (*SOX2 and KLF4*) in human cells (Zhang et al. 2011). Sanjana et al. proposed a protocol that could facilitate the rapid construction of TALENs within one week and could be easily scaled up to construct TALEs for multiple targets in parallel (Sanjana et al. 2012). As the earliest GG-PCR method to construct TALE repeats, this protocol successfully took advantage of the degeneracy in the codons encoding glycine–leucine residues at the junction of two adjacent TALE units. The design of PCR primers considered the production of 24 different codons for the sticky ends between different TALE repeat units and introduced the recognition sites of type IIS endonuclease. In this manner, the cleavage of the PCR products of different TALE repeat units by type IIS endonuclease resulted in fragments with 5′-protruding sticky ends at both ends. Subsequently, 4 or 6 TALE units were sequentially connected through different sticky ends to form a fragment containing 4 or 6 TALE repeat units. Further, multiple 4- or 6-unit fragments were sequentially connected with the same method to obtain TALEs with the required length, which were then cloned into the vector fused with Fok I to complete TALEN vector construction. Later, several other laboratories also successfully constructed TALEN expression vectors using the GG-Vector method. Geiβler et al. created synthetic TAL proteins with designed repeat compositions using a novel modular cloning strategy termed "Golden TAL Technology." These new programmed TAL proteins were functional not only in plant cells but also in human cells and activated targeted expression of exogenous as well as endogenous genes. Transcriptional activation in different human cell lines was significantly improved by replacing the TAL-AD with the VP16-AD of herpes simplex virus (Geissler et al. 2011). Morbitzer et al. used type IIS restriction enzymes in two sequential cut-ligase reactions to construct dTALE repeat arrays. The dTALE generated by this modular approach could differentiate between two highly similar DNA sequences that were both targeted by the *Xanthomonas* TALE AvrBs3. These data showed that this modular assembly system allows the rapid generation of highly specific TALE-type DNA-binding domains that target binding sites of predefined length and sequence. This approach enables the rapid and flexible production of dTALEs for gene regulation and genome editing in routine and high-throughput applications (Morbitzer et al. 2011). Weber et al. engineered TALE proteins with novel DNA-binding specificities based on the 17.5 repeat-containing AvrBs3 TALE as a scaffold. For each of the 17 full repeats, four module types were generated, each with a distinct base preference. Using this set of 68 repeat modules, recognition domains for any 17-nucleotide DNA target sequence of choice could be constructed by assembling selected modules in a defined linear order. Assembly was conducted in two successive cloning steps using the Golden Gate cloning method that allows the seamless fusion of multiple DNA fragments (Weber et al. 2011).

Basically, the abovementioned method constructs four basic TALE repeat units with eight different sticky ends and clones them into a vector as basic materials for construction. To assemble TALEN, the corresponding vectors are selected for enzyme digestion and ligation to first assemble a vector containing multiple TALE repeat units. Subsequently, the ligated TALE repeat units are further cut and ligated to assemble the desired TALE sequence, which is cloned into a vector fused with Fok I.

3) The restriction enzyme and ligation (REAL) method

The REAL method, invented in the laboratory of Sander et al., is an assembly method based on repeated digestion–ligation–transformation–screening (Sander et al. 2011). In addition to adding different sticky ends and the recognition sites of type IIS endonuclease Bbs I and Bsa I at the 5′ and 3′ ends of the four basic TALE repeat units, this method adds a recognition site of common restriction endonuclease BamH I to the downstream of the Bsa I recognition site. The fragments are then cloned into a suitable vector for assembly. During assembly, the vector that contains the first TALE unit is subjected to Bsa I and BamH I double enzyme digestion, resulting in a 4-bp specific sticky end and a BamH I sticky end on the vector. Meanwhile, Bbs I and BamH I double enzyme digestion of the vector containing the next TALE unit generates the fragment with a 4-bp specific sticky end at the 5′ end and a BamH I sticky end at the 3′ end. This fragment is complementary to the two sticky ends of the vector where the first unit is located; therefore, the fragment can be ligated into the vector transforming bacteria to obtain the assembled vector containing the dual-TALE units. The newly obtained vector still contains the recognition sites of type IIS endonuclease Bbs I, Bsa I, and restriction endonuclease BamH I, allowing the repetition of the similar digestion–ligation–transformation process until all TALE units are ligated into the expression vector to construct the TALE repeat sequence required for target recognition.

Considering the method above is too cumbersome because all TALE units have to be ligated one by one, the REAL-Fast method was introduced to reduce the number of rounds of the enzyme digestion–ligation–transformation cycle by using the vectors containing multi-TALE units as basic materials. Using the method described above, a total of 384 kinds of vectors containing different dual-TALE units, triple-TALE units, and quadruple-TALE units can be preassembled (Reyon et al. 2012).

4) The unit assembly method

The unit assembly method is another enzyme digestion–ligation-based method for assembly of TALE repeat domains developed by Huang et al. (Huang et al.

2011). TALE repeat sequences are head-to-tail tandem repeats with differences in only the 12th and 13th amino acids. Hence, the unit assembly method uses the fragment between serine (Ser) at position 11 of a TALE repeat unit and alanine (Ala) at position 10 of the next repeat unit as a new repeat unit (also known as "substituted unit"). At the 5′ and 3′ ends of this unit, the recognition site of type IIS endonuclease is substituted with the recognition sites of restriction endonuclease Spe I and its isocaudomer Nhe I, and a Hind III site is added downstream of the enzyme digestion site of Nhe I. Four basic TALE substituted units with the above new enzyme recognition sites are cloned into suitable vectors. To construct a TALE expression vector, the vector containing the first substituted unit can be cut by Nhe I and Hind III, and then the fragment of the next substituted unit can be obtained by Spe I and Hind III digestion. Spe I and Nhe I are a pair of isocaudomers whose digestion creates the same sticky ends; therefore, the sticky ends resulting from both enzymes can be ligated with the sticky end produced by Hind III to integrate the next substituted unit fragment into the vector containing the first substituted unit. Subsequently, the obtained vector is ligated and transformed into bacteria to construct the vector containing the dual-TALE substitution unit. Through this process, the constructed vector still contains the recognition sites of Spe I, Nhe I, and Hind III, allowing the same digestion–ligation–transformation process to repeat until all TALE substitution units are successively ligated into the vector. Finally, the obtained TALE repeat sequence is cleaved out by Spe I and Nhe I and cloned into the vector fused with Fok I to complete the construction of the TALEN expression vector. It is worth noting that this method uses isocaudomers instead of type IIS endonuclease so that the sticky ends created by enzymatic digestion are the same between the two TALE substituted units. Hence, the materials required for construction are very simple: only four vectors containing basic TALE substituted units are needed for subsequent expression vector construction. In addition, Spe I, Nhe I, and Hind III are common restriction enzymes that are much less expensive than type IIS endonuclease. Therefore, this method greatly reduces the cost of vector construction and is ideal for most general laboratories. Similar to the REAL-Fast method, this method can also preassemble 16 kinds of vectors containing dual-TALE unit, 64 kinds of triple-TALE unit, and 256 kinds of quadruple-TALE unit, making the subsequent construction easier and faster. To date, the TALEN expression vector constructed by Zhang Bo's laboratory using this method has successfully achieved genomic site-directed mutagenesis and gene knockout in zebrafish (Zu et al. 2013).

2.4.3.3 Construction of CRISPR/Cas9 Expression Vector

Efficient plant CRISPR/Cas9-based genome editing usually includes four steps (Figure 2.3). Step 1 is to design and construct gene-specific sgRNA. Various online tools have been developed for computer-aided sgRNA design. However, these tools are not fully applicable in plants, and their computing accuracy for sgRNA

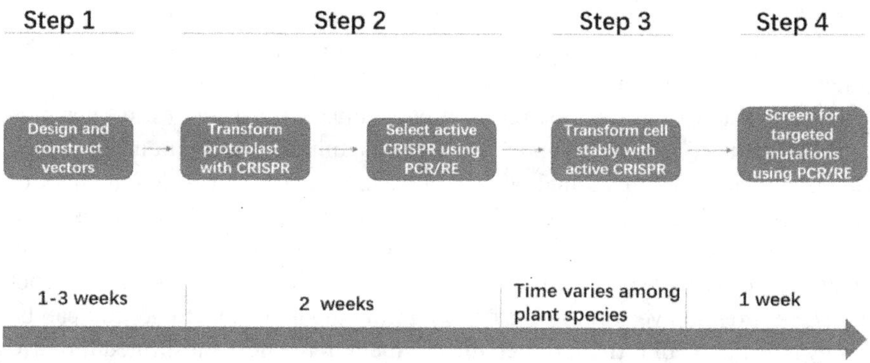

Figure 2.3 General procedures of CRISPR/Cas9 for plant genome editing.

selection needs improvement, requiring large-scale data collection and systematic research on sgRNA efficiency in plant cells. Step 2 is to verify the activity of sgRNA in protoplasts prior to its use for genome editing. Step 3 is to deliver the components of the CRISPR/Cas9 system into plant cells by *Agrobacterium*-mediated transformation or particle bombardment and stably integrate the Cas9 and sgRNA expression cassette into the plant genome. Finally, Step 4 is to verify the required modifications by PCR genotyping of the transformed/regenerated plants and further confirm the results by sequencing.

The CRISPR/Cas9 system is an efficient and rapid tool for gene editing and can provide a powerful platform for basic research; however, further research has revealed its major drawback of off-target effects. Therefore, the factors involved in the off-target effects of the CRISPR/Cas9 gene-editing system and the strategies for solving the related issues have become hotspots in the research of this technology.

The specificity of a CRISPR/Cas9 gene-editing system depends on the recognition sequence (~20 nt) of sgRNA. However, the recognition sequence of sgRNA may partially pair nontarget DNA in the complex biological genomes. There are three prominent factors affecting the off-target effects of a CRISPR/Cas9 system. The first factor is PAM. The presence of a PAM sequence is the premise for targeted cleavage by a CRISPR/Cas9 system. Without the presence of a PAM sequence at the 3′ end of the target sequence, Cas9 will not cleave the sequence site even when the target sequence completely matches the sgRNA sequence (Zhang et al. 2014). The PAM sequence mainly affects the DNA cleavage efficiency of CRISPR/Cas9. The second factor is the sgRNA sequence. A 20-nt guide sequence within the sgRNA directs Cas9 to the genomic target, which determines the targeting specificity of the CRISPR/Cas system. However, some studies argued that the specificity of CRISPR/Cas9 and target site recognition mainly depends on the pairing of sgRNA with the 10–12 bases close to the PAM region, while mismatching of the other 8–10 bases away from the PAM sequence does not significantly affect target site recognition (Jinek et al. 2012; Cong et al. 2013; Jiang et al. 2013). The third specificity-related factor is the length of sgRNA. Studies have shown that shortening of the sgRNA target sequence from 20 nt to 17 nt can greatly reduce the off-target effects without significantly affecting the cleavage efficiency (Fu et al. 2014). However, Cho et al. stated that the addition of two guanine nucleotides to the 5′ end of the designed sgRNA can significantly improve the specificity of CRISPR/Cas9 system (Cho et al. 2014). The two approaches seem to contradict each other, but in essence, they act on different aspects to improve the binding specificity of sgRNA.

2.4.4 Application

Rapid and efficient plant genome-editing technology is a powerful tool for basic research and crop improvement. Osakabe et al. used ZFN to inactivate the endogenous genes of interest in *Arabidopsis thaliana*. ZFN specifically cleaved its recognition sequence (*ABI4* gene) in vitro. In *Arabidopsis* where the ZFN gene driven by the heat shock promoter was introduced into the genome, ZFN-mediated cleavage induced by heat shock resulted in gene mutations with deletion and substitution in the *ABI4* gene, achieving a gene mutation efficiency of up to 3%. In addition, homozygous *ABI4* mutants clearly exhibited the expected ABA- and glucose-insensitive phenotype (Osakabe et al. 2010). Hilioti et al. developed a ZFN-based technology for *Solanum lycopersicum* breeding. ZFNs designed for the developmental regulator *LEAFY-COTYLEDON1-LIKE4* (*L1L4*) coding for the β subunit of nuclear factor Y, when transiently expressed in tomato seeds, cleaved the target site and stimulated NHEJ-driven incomplete repair to mutate the endogenous target site, thus, introducing *LIL4* mutations. The results showed that sequence changes upstream of the DNA-binding domain of *L1L4* can lead to phenotypic diversity, including of the fruit organ (Hilioti et al. 2016).

Budhagatapalli et al. pioneered a new approach for targeted mutagenesis in eukaryotic genomes using TALENs. They used *gfp*-specific TALENs, along with an exogenous repair template that facilitates conversion of *gfp* into *yfp*, to promote a single amino acid exchange in the gene product via HDR. The results showed that yellow fluorescent protein accumulation was detected in approximately 100 mutated cells, and the creation of a functional *yfp* gene via HDR was unambiguously confirmed by sequencing of the respective genomic site (Budhagatapalli et al. 2015). Blanvillain-Baufume et al. generated an allele library of the *OsSWEET14* promoter through stable expression of TALEN constructs in rice and found that plants edited in AvrXa7 or Tal5 EBEs were resistant to bacterial strains relying on the corresponding TALE (Blanvillain-Baufume et al. 2017). Ma et al. performed targeted mutagenesis in rice using TALENs to improve seed storability and found that one heterozygous mutant (4%) was recovered from 25 transgenic

NPTII-resistant lines, and the mutation was transmitted to the next generation. Through further molecular- and protein-level experiments, they verified LOX3 deficiency and confirmed the improvement in seed storability (Ma et al. 2015).

Hyun et al. used the RNA-guided endonuclease (RGEN) derived from the CRISPR/Cas9 protein system for site-directed mutagenesis of the *Arabidopsis* genome. *U6 snRNA* and *INCURVATA2* promoters were used to express sgRNA and Cas9 DNA endonuclease, respectively. After *Agrobacterium*-mediated introduction of T-DNAs encoding RGENs that target *FLOWERING LOCUS T* (*FT*) and *SQUAMOSA PROMOTER BINDING PROTEIN-LIKE 4* genes, somatic mutagenesis at the targeted loci was observed in T1 transformants. In the results of FT-RGEN, T1 plants often demonstrated late flowering. DNA sequencing analysis revealed that approximately 90% of independent chromosomal DNA fragments carried mutations in the analyzed tissue of a T1 plant showing late flowering (Hyun et al. 2015). Ito et al. performed efficient mutagenesis of the tomato genome by using the CRISPR/Cas9 system targeting the *RIN* gene, which encodes a MADS-box transcription factor that regulates tomato ripening. The results showed that three regions within the gene were targeted, and mutations consisting either of a single-base insertion or deletion of more than three bases were found at the Cas9 cleavage sites in T_0 regenerated plants. The RIN-protein-defective mutants produced incomplete-ripening fruits in which the enrichment of red color pigments (mainly lycopene and β-carotene) was significantly lower than that of the wild type (Ito et al. 2015). Jia et al. used CRISPR/Cas9/sgRNA technology to modify the canker susceptibility gene *CsLOB1* in Duncan grapefruit and create a citrus variety with significantly enhanced disease resistance, providing a promising pathway to generate disease-resistant citrus varieties (Jia et al. 2017).

Rodriguez-Leal et al. (2017) used a CRISPR/Cas9 system targeting gene promoter regions to create different cis-regulatory mutations and alleles for rapid evaluation of the phenotypic impact of numerous promoter variants for genes regulating three major productivity traits in tomato—fruit size, inflorescence branching, and plant architecture—which provides useful quantitative variation for crop improvement.

Recently, Li et al. generated knockout *cr-rz1al* mutant lines via the CRISPR/Cas9 gene-editing system to analyze the functions of *RZ1AL* in fruit development and ripening. Knockout of *RZ1AL* reduced fruit lycopene content and weight in the *cr-rz1al* mutant plants. (Figure 2.4) RNA-Seq data demonstrated that the expression levels of genes that encode several key enzymes associated with carotenoid biosynthesis and metabolism were notably downregulated in *cr-rz1al* fruits. At the same time, Proteomic analysis revealed that the levels of various ribosomal subunit proteins were reduced. This could affect the translation of ripening-related proteins such as ZDS. Collectively, our findings demonstrate that *RZ1AL* may participate in the regulation of carotenoid biosynthesis and metabolism and affect tomato development and fruit ripening (Li et al. 2022).

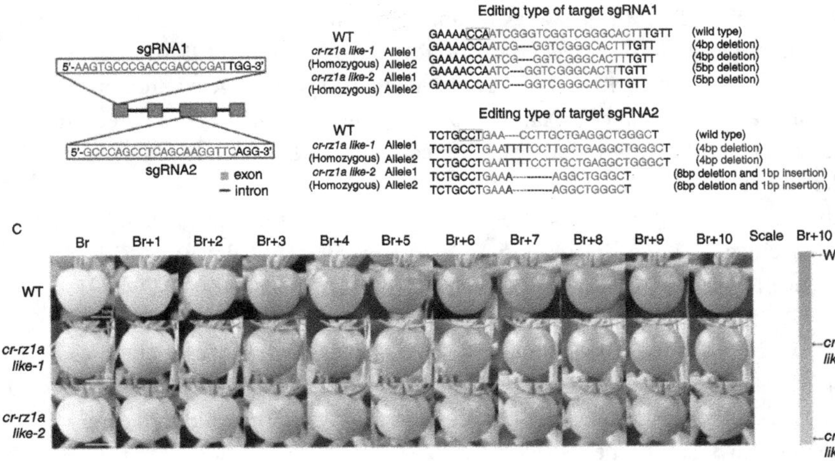

Figure 2.4 Application of CRISPR/Cas9 gene-editing system (Li et al. 2022/from Oxford University Press).

References

Ai, T.N., Naing, A.H., Arun, M. et al. (2016). Sucrose-induced anthocyanin accumulation in vegetative tissue of Petunia plants requires anthocyanin regulatory transcription factors. *Plant Sci* 252: 144–150.

Aigner, T., Haag, J., and Zimmer, R. (2007). Functional genomics, evo-devo and systems biology: a chance to overcome complexity? *Curr Opin Rheumatol* 19 (5): 463–470.

Angenent, G.C., Linthorst, H.J., van Belkum, A.F. et al. (1986). RNA 2 of tobacco rattle virus strain TCM encodes an unexpected gene. *Nucl Acids Res* 14: 4673–4682.

Baliga, N.S. (2008). Systems biology. The scale of prediction. *Science* 320 (5881): 1297–1298.

Barrangou, R., Fremaux, C., Deveau, H. et al. (2007). CRISPR provides acquired resistance against viruses in prokaryotes. *Science* 315: 1709–1712. doi:10.1126/science.1138140.

Baulcombe, D. (2004). RNA silencing in plants. *Nature* 431: 356–363.

Beloglazova, N., Brown, G., Zimmerman, M.D. et al. (2008). A novel family of sequence-specific endoribonucleases associated with the clustered regularly interspaced short palindromic repeats. *J Biol Chem* 283: 20361–20371. doi:10.1074/jbc.M803225200.

Benedito, V.A., Visser, P.B., Angenent, G.C., and Krens, F.A. (2004). The potential of virus-induced gene silencing for speeding up functional characterization of plant genes. *Genet Mol Res* 3: 323–341.

Bernstein, E., Caudy, A.A., Hammond, S.M., and Hannon, G.J. (2001). Role for a bidentate ribonuclease in the initiation step of RNA interference. *Nature* 409: 363–366.

Bitinaite, J., Wah, D.A., Aggarwal, A.K., and Schildkraut, I. (1998). FokI dimerization is required for DNA cleavage. *Proc Natl Acad Sci USA* 95 (18): 10570–10575.

Blanvillain-Baufume, S., Reschke, M., Sole, M. et al. (2017). Targeted promoter editing for rice resistance to Xanthomonas oryzae pv. oryzae reveals differential activities for SWEET14-inducing TAL effectors. *Plant Biotechnol J* 15: 306–317. doi:10.1111/pbi.12613.

Boch, J., Scholze, H., Schornack, S. et al. (2009). Breaking the code of DNA binding specificity of TAL-type III effectors. *Science* 326: 1509–1512. doi:10.1126/science.1178811.

Bolotin, A., Quinquis, B., Sorokin, A., and Ehrlich, S.D. (2005). Clustered regularly interspaced short palindrome repeats (CRISPRs) have spacers of extrachromosomal origin. *Microbiology+* 151: 2551–2561. doi:10.1099/mic.0.28048-0.

Brigneti, G., Martín Hernández, A.M., Jin, H. et al. (2004). Virus-induced gene silencing in Solanum species. *Plant J* 39: 264–272.

Brouns, S.J., Jore, M.M., Lundgren, M. et al. (2008). Small CRISPR RNAs guide antiviral defense in prokaryotes. *Science* 321: 960–964. doi:10.1126/science.1159689.

Budhagatapalli, N., Rutten, T., Gurushidze, M. et al. (2015). Targeted modification of gene function exploiting homology-directed repair of TALEN-mediated double-strand breaks in Barley. *G3 (Bethesda)* 5: 1857–1863. doi:10.1534/g3.115.018762.

Burch Smith, T.M., Anderson, J.C., Martin, G.B., and Dinesh Kumar, S.P. (2004). Applications and advantages of virus-induced gene silencing for gene function studies in plants. *Plant J* 39: 734–746.

Cermak, T., Doyle, E.L., Christian, M. et al. (2011). Efficient design and assembly of custom TALEN and other TAL effector-based constructs for DNA targeting. *Nucleic Acids Res* 39: e82. doi:10.1093/nar/gkr218.

Chai, Y., Jia, H., Li, C. et al. (2011). FaPYR1 is involved in strawberry fruit ripening. *J Exp Bot* 62: 5079–5089.

Chanderbali, A.S., Albert, V.A., Leebens-Mack, J. et al. (2009). Transcriptional signatures of ancient floral developmental genetics in avocado (Persea americana; Lauraceae). *Proc Natl Acad Sci USA* 106: 8929–8934.

Chen, K.C., Calzone, L., Csikasz-Nagy, A. et al. (2004). Integrative analysis of cell cycle control in budding yeast. *Mol Biol Cell* 15 (8): 3841–3862.

Cho, S.W., Kim, S., Kim, Y. et al. (2014). Analysis of off-target effects of CRISPR/Cas-derived RNA-guided endonucleases and nickases. *Genome Res* 24: 132–141. doi:10.1101/gr.162339.113.

Chung, E., Seong, E., Kim, Y.C. et al. (2004). A method of high frequency virus induced gene silencing in chili pepper (Capsicum annuum L. cv. Bukang). *Mol Cells* 17: 377–380.

Cong, L., Ran, F.A., Cox, D. et al. (2013). Multiplex genome engineering using CRISPR/Cas systems. *Science* 339: 819–823. doi:10.1126/science.1231143.

Constantin, G.D., Krath, B.N., MacFarlane, S.A. et al. (2004). Virus-induced gene silencing as a tool for functional genomics in a legume species. *Plant J* 40: 622–631.

Cui, H. and Wang, A. (2017). An efficient viral vector for functional genomic studies of Prunus fruit trees and its induced resistance to Plum pox virus via silencing of a host factor gene. *Plant Biotechnol J* 15: 344–356.

Ding, X.S., Schneider, W.L., Chaluvadi, S.R. et al. (2006). Characterization of a Brome mosaic virus strain and its use as a vector for gene silencing in monocotyledonous hosts. *Mol Plant-Microbe Interactions* 19: 1229–1239.

Du, M., Zhai, Q., Deng, L. et al. (2014). Closely related NAC transcription factors of tomato differentially regulate stomatal closure and reopening during pathogen attack. *Plant Cell* 26 (7): 3167–3184.

Elitzur, T., Yakir, E., Quansah, L. et al. (2016). Banana MaMADS transcription factors are necessary for fruit ripening and molecular tools to promote shelf-life and food security. *Plant Physiol* 171 (1): 380–391.

Faivre-Rampant, O., Gilroy, E.M., Hrubikova, K. et al. (2004). Potato virus X-induced gene silencing in leaves and tubers of potato. *Plant Physiol* 134: 1308–1316.

Fantini, E. and Giuliano, G. (2016). Virus-induced gene silencing as a tool to study tomato fruit biochemistry. *Methods Mol Biol (Clifton, N.J.)* 1363: 65–78.

Fofana, I.B., Sangare, A., Collier, R. et al. (2004). A geminivirus-induced gene silencing system for gene function validation in cassava. *Plant Mol Biol* 56: 613–624.

Fu, Y., Foden, J.A., Khayter, C. et al. (2013). High-frequency off-target mutagenesis induced by CRISPR-Cas nucleases in human cells. *Nat Biotechnol* 31: 822–826. doi:10.1038/nbt.2623.

Fu, Y., Sander, J.D., Reyon, D. et al. (2014). Improving CRISPR-Cas nuclease specificity using truncated guide RNAs. *Nat Biotechnol* 32: 279–284. doi:10.1038/nbt.2808.

Fukuoka, Y., Lindgren, T.G., Rankin, S.H. et al. (2007). Cluster analysis: a useful technique to identify elderly cardiac patients at risk for poor quality of life. *Qual Life Res* 16 (10): 1655–1663.

Geissler, R., Scholze, H., Hahn, S. et al. (2011). Transcriptional activators of human genes with programmable DNA-specificity. *PLoS One* 6: e19509. doi:10.1371/journal.pone.0019509.

Gosselé, V., Faché, I., Meulewaeter, F. et al. (2002). SVISS–a novel transient gene silencing system for gene function discovery and validation in tobacco plants. *Plant J* 32: 859–866.

Gray, J., Picton, S., Shabbeer, J. et al. (1992). Molecular biology of fruit ripening and its manipulation with antisense genes. *Plant Mol Biol* 19 (1): 69–87.

Hammond, S.M., Bernstein, E., Beach, D., and Hannon, G.J. (2000). An RNA-directed nuclease mediates post-transcriptional gene silencing in Drosophila cells. *Nature* 404: 293–296.

Han, X., Feng, Z., Xing, D. et al. (2015). Two NAC transcription factors from Caragana intermedia altered salt tolerance of the transgenic Arabidopsis. *BMC Plant Boil* 15: 208.

Herrera-Estrella, L., Simpson, J., and Martínez-Trujillo, M. (2004). Transgenic plants: an historical perspective. *Methods Mol Biol (Clifton, N.J.)* 286: 3–32.

Hilioti, Z., Ganopoulos, I., Ajith, S. et al. (2016). A novel arrangement of zinc finger nuclease system for in vivo targeted genome engineering: the tomato LEC1-LIKE4 gene case. *Plant Cell Rep* 35: 2241–2255. doi:10.1007/s00299-016-2031-x.

Holzberg, S., Brosio, P., Gross, C., and Pogue, G.P. (2002). Barley stripe mosaic virus-induced gene silencing in a monocot plant. *Plant J* 30: 315–327.

Hsu, P.D., Scott, D.A., Weinstein, J.A. et al. (2013). DNA targeting specificity of RNA-guided Cas9 nucleases. *Nat Biotechnol* 31: 827–832. doi:10.1038/nbt.2647.

Huang, P., Xiao, A., Zhou, M. et al. (2011). Heritable gene targeting in zebrafish using customized TALENs. *Nat Biotechnol* 29: 699–700. doi:10.1038/nbt.1939.

Hyun, Y., Kim, J., Cho, S.W. et al. (2015). Site-directed mutagenesis in Arabidopsis thaliana using dividing tissue-targeted RGEN of the CRISPR/Cas system to generate heritable null alleles. *Planta* 241: 271–284. doi:10.1007/s00425-014-2180-5.

Igarashi, A., Yamagata, K., Sugai, T. et al. (2009). Apple latent spherical virus vectors for reliable and effective virus-induced gene silencing among a broad range of plants including tobacco, tomato, Arabidopsis thaliana, cucurbits, and legumes. *Virology* 386: 407–416.

Ito, Y., Nishizawa-Yokoi, A., Endo, M. et al. (2015). CRISPR/Cas9-mediated mutagenesis of the RIN locus that regulates tomato fruit ripening. *Biochem Biophys Res Commun* 467: 76–82. doi:10.1016/j.bbrc.2015.09.117.

Jaakola, L., Poole, M., Jones, M.O. et al. (2010). A SQUAMOSA MADS box gene involved in the regulation of anthocyanin accumulation in bilberry fruits. *Plant Physiol* 153: 1619–1629.

Jia, H., Chai, Y., Li, C. et al. (2011). Abscisic acid plays an important role in the regulation of strawberry fruit ripening. *Plant Physiol* 157: 188–199.

Jia, H., Zhang, Y., Orbovic, V. et al. (2017). Genome editing of the disease susceptibility gene CsLOB1 in citrus confers resistance to citrus canker. *Plant Biotechnol J* 15: 817–823. doi:10.1111/pbi.12677.

Jiang, W., Bikard, D., Cox, D. et al. (2013). RNA-guided editing of bacterial genomes using CRISPR-Cas systems. *Nat Biotechnol* 31: 233–239. doi:10.1038/nbt.2508.

Jinek, M., Chylinski, K., Fonfara, I. et al. (2012). A programmable dual-RNA-guided DNA endonuclease in adaptive bacterial immunity. *Science* 337: 816–821. doi:10.1126/science.1225829.

Kayser, K., Kayser, G., Metze, K. et al. (2007). The concept of structural entropy in tissue-based diagnosis. *Anal Quant Cytol Histol* 29 (5): 296–308.

Kikuchi, Y., Hijikata, N., Ohtomo, R. et al. (2016). Aquaporin-mediated long-distance polyphosphate translocation directed towards the host in arbuscular mycorrhizal symbiosis: application of virus-induced gene silencing. *New Phytologist* 211: 1202–1208.

Kim, J., Park, M., Jeong, E.S. et al. (2017). Harnessing anthocyanin-rich fruit: a visible reporter for tracing virus-induced gene silencing in pepper fruit. *Plant Methods* 13: 3.

Kim, S., Lee, M.J., Kim, H. et al. (2011). Preassembled zinc-finger arrays for rapid construction of ZFNs. *Nat Methods* 8: 7. doi:10.1038/nmeth0111-7a.

Kitano, H. (2007). The theory of biological robustness and its implication in cancer. *Ernst Schering Res Found Workshop* 61: 69–88.

Kjemtrup, S., Sampson, K.S., Peele, C.G. et al. (1998). Gene silencing from plant DNA carried by a geminivirus. *Plant J* 14: 91–100.

Klug, A. (2010). The discovery of zinc fingers and their development for practical applications in gene regulation and genome manipulation. *Q Rev Biophys* 43: 1–21. doi:10.1017/S0033583510000089.

Koschorreck, M., Conzelmann, H., Ebert, S. et al. (2007). Reduced modeling of signal transduction– a modular approach. *BMC Bioinformatics* 8: 336.

Kumagai, M.H., Donson, J., Della-Cioppa, G. et al. (1995). Cytoplasmic inhibition of carotenoid biosynthesis with virus-derived RNA. *Proc Natl Acad Sci USA* 92: 1679–1683.

Li, S., Xu, H., Ju, Z. et al. (2018). The RIN-MC fusion of MADS-Box transcription factors has transcriptional activity and modulates expression of many ripening genes. *Plant Physiol* 176 (1): 891–909.

Li, T., Huang, S., Zhao, X. et al. (2011). Modularly assembled designer TAL effector nucleases for targeted gene knockout and gene replacement in eukaryotes. *Nucleic Acids Res* 39: 6315–6325. doi:10.1093/nar/gkr188.

Li, X., Yang, Y., Zeng, N. et al. (2022). Glycine-rich RNA-binding cofactor RZ1AL is associated with tomato ripening and development. *Hortic Res* 9: uhac134.

Lijoi, A., Mena, R.H., Prünster, I. et al. (2007). A Bayesian nonparametric method for prediction in EST analysis. *BMC Bioinformatics* 8: 339.

Lindbo, J.A., Silva-Rosales, L., Proebsting, W.M., and Dougherty, W.G. (1993). Induction of a highly specific antiviral state in transgenic plants: implications for regulation of gene expression and virus resistance. *Plant Cell* 5: 1749–1759.

Liu, G., Li, H., and Fu, D. (2021). Applications of virus-induced gene silencing for identification of gene function in fruit. *Food Quality Safety* 5: fyab018.

Liu, N., Xie, K., Jia, Q. et al. (2016). Foxtail mosaic virus-induced gene silencing in monocot plants. *Plant Physiol* 171 (3): 1801–1807.

Liu, Y., Schiff, M., Marathe, R., and Dinesh Kumar, S.P. (2002c). Tobacco Rar1, EDS1 and NPR1/NIM1 like genes are required for N-mediated resistance to tobacco mosaic virus. *Plant J* 30: 415–429.

Lu, H., Chen, H., Tsai, W. et al. (2007). Strategies for functional validation of genes involved in reproductive stages of orchids. *Plant Physiol* 143: 558–569.

Lu, R. (2002). High throughput virus induced gene silencing for the analysis of disease resistance in plants. *J Pak Med Assoc* 52: 459–465.

Ma, L., Yang, Y., Wang, Y. et al. (2022). SlRBP1 promotes translational efficiency via SleIF4A2 to maintain chloroplast function in tomato. *Plant Cell* 34 (7): 2747–2764.

Ma, L., Zhu, F., Li, Z. et al. (2015). TALEN-based mutagenesis of lipoxygenase LOX3 enhances the storage tolerance of rice (Oryza sativa) seeds. *PLoS One* 10: e0143877. doi:10.1371/journal.pone.0143877.

MacFarlane, S.A. (1999). Molecular biology of the tobraviruses. *J Gen Virol* 80: 2799–2807.

Maeder, M.L., Thibodeau-Beganny, S., Osiak, A. et al. (2008). Rapid "open-source" engineering of customized zinc-finger nucleases for highly efficient gene modification. *Mol Cell* 31: 294–301.

Makarova, K.S., Haft, D.H., Barrangou, R. et al. (2011). Evolution and classification of the CRISPR-Cas systems. *Nat Rev Microbiol* 9: 467. doi.

Mei, Y., Zhang, C., Kernodle, B.M. et al. (2016). A Foxtail mosaic virus vector for virus-induced gene silencing in maize. *Plant Physiol* 171: 760–772.

Meselson, M. and Stahl, F.W. (1958). The replication of DNA in Escherichia coli. *Proc Natl Acad Sci U S A*. 15; 44 (7): 671–682.

Miller, J.C., Tan, S., Qiao, G. et al. (2011). A TALE nuclease architecture for efficient genome editing. *Nat Biotechnol* 29: 143–148. doi:10.1038/nbt.1755.

Mojica, F.J., Diez-Villasenor, C., Garcia-Martinez, J., and Almendros, C. (2009). Short motif sequences determine the targets of the prokaryotic CRISPR defence system. *Microbiology+* 155: 733–740. doi:10.1099/mic.0.023960-0.

Mojica, F.J., Diez-Villasenor, C., Garcia-Martinez, J., and Soria, E. (2005). Intervening sequences of regularly spaced prokaryotic repeats derive from foreign genetic elements. *J Mol Evol* 60: 174–182. doi:10.1007/s00239-004-0046-3.

Morbitzer, R., Elsaesser, J., Hausner, J., and Lahaye, T. (2011). Assembly of custom TALE-type DNA binding domains by modular cloning. *Nucleic Acids Res* 39 (13): 5790–5799.

Moscou, M.J. and Bogdanove, A.J. (2009). A simple cipher governs DNA recognition by TAL effectors. *Science (New York, N.Y.)* 326 (5959): 1501.

Muruganantham, M., Moskovitz, Y., Haviv, S. et al. (2009). Grapevine virusA-mediated gene silencing in Nicotiana benthamiana and Vitis vinifera. *J Virological Methods* 155: 167–174.

Mussolino, C., Morbitzer, R., Lutge, F. et al. (2011). A novel TALE nuclease scaffold enables high genome editing activity in combination with low toxicity. *Nucleic Acids Res* 39: 9283–9293. doi:10.1093/nar/gkr597.

Naylor, M., Reeves, J., Cooper, J.I. et al. (2005). Construction and properties of a gene-silencing vector based on Poplar mosaic virus (genus Carlavirus). *J Virological Methods* 124: 27–36.

Negi, S., Tak, H., and Ganapathi, T.R. (2016). Functional characterization of secondary wall deposition regulating transcription factors MusaVND2 and MusaVND3 in transgenic banana plants. *Protoplasma* 253 (2): 431–446.

Nguyen, C.V., Vrebalov, J.T., Gapper, N.E. et al. (2014). Tomato GOLDEN2-LIKE transcription factors reveal molecular gradients that function during fruit development and ripening. *Plant Cell* 26 (2): 585–601.

Nikolov, S., Yankulova, E., Wolkenhauer, O. et al. (2007). Principal difference between stability and structural stability (robustness) as used in systems biology. *Nonlinear Dynamics Psychol Life Sci* 11 (4): 413–433.

Nirenberg, M.W., Matthaei, J.H., Jones, O.W. et al. (1963). Approximation of genetic code via cell-free protein synthesis directed by template RNA. *Fed Proc* 22: 55–61. PMID: 13938750.

Nishikura, K. (2001). A short primer on RNAi: RNA-directed RNA polymerase acts as a key catalyst. *Cell* 107: 415–418.

Nykänen, A., Haley, B., and Zamore, P.D. (2001). ATP requirements and small interfering RNA structure in the RNA interference pathway. *Cell* 107: 309–321.

Osakabe, K., Osakabe, Y., and Toki, S. (2010). Site-directed mutagenesis in Arabidopsis using custom-designed zinc finger nucleases. *Proc Natl Acad Sci U S A* 107: 12034–12039. doi:10.1073/pnas.1000234107.

Pacak, A., Geisler, K., Jørgensen, B. et al. (2010). Investigations of barley stripe mosaic virus as a gene silencing vector in barley roots and in Brachypodium distachyon and oat. *Plant Methods* 6: 26.

Palmer, K.E. and Rybicki, E.P. (2001). Investigation of the potential of maize streak virus to act as an infectious gene vector in maize plants. *Arch Virol* 146: 1089–1104.

Pan, X.Q., Fu, D.Q., Zhu, B.Z. et al. (2013b). Overexpression of the ethylene response factor SlERF1 gene enhances resistance of tomato fruit to Rhizopus nigricans. *Postharvest Biol Technol* 75: 28–36.

Peele, C., Jordan, C.V., Muangsan, N. et al. (2001). Silencing of a meristematic gene using geminivirus-derived vectors. *Plant J* 27: 357–366.

Porteus, M.H. and Carroll, D. (2005). Gene targeting using zinc finger nucleases. *Nat Biotechnol* 23: 967–973. doi:10.1038/nbt1125.

Pourcel, C., Salvignol, G., and Vergnaud, G. (2005). CRISPR elements in Yersinia pestis acquire new repeats by preferential uptake of bacteriophage DNA, and provide additional tools for evolutionary studies. *Microbiology+* 151: 653–663. doi:10.1099/mic.0.27437-0.

Purkayastha, A. and Dasgupta, I. (2009). Virus-induced gene silencing: a versatile tool for discovery of gene functions in plants. *Plant Physiol Biochem* 47: 967–976.

Qi, L.S., Larson, M.H., Gilbert, L.A. et al. (2013). Repurposing CRISPR as an RNA-guided platform for sequence-specific control of gene expression. *Cell* 152: 1173–1183. doi:10.1016/j.cell.2013.02.022.

Ratcliff, F., Martin Hernandez, A.M., and Baulcombe, D.C. (2001). Technical advance: tobacco rattle virus as a vector for analysis of gene function by silencing. *Plant J* 25: 237–245.

Renner, T., Bragg, J., Driscoll, H.E. et al. (2009). Virus-induced gene silencing in the culinary ginger (Zingiber officinale): an effective mechanism for down-regulating gene expression in tropical monocots. *Molecular Plant* 2: 1084–1094.

Reyon, D., Tsai, S.Q., Khayter, C. et al. (2012). FLASH assembly of TALENs for high-throughput genome editing. *Nat Biotechnol* 30: 460–465. doi:10.1038/nbt.2170.

Richmond, T. and Somerville, S. (2000). Chasing the dream: plant EST microarrays. *Curr Opinion Plant Biol* 3: 108–116.

Robertson, D. (2004). VIGS vectors for gene silencing: many targets, many tools. *Annu Rev Plant Biol* 55: 495–519.

Rodriguez-Leal, D., Lemmon, Z.H., Man, J. et al. (2017). Engineering Quantitative Trait Variation for Crop Improvement by Genome Editing. *Cell* 171: 470–480.e8. doi:10.1016/j.cell.2017.08.030.

Rössner, C., Lotz, D., and Becker, A. (2022). VIGS Goes Viral: how VIGS Transforms Our Understanding of Plant Science. *Annu Rev Plant Biol* 73: 703–728.

Ruiz, M.T., Voinnet, O., and Baulcombe, D.C. (1998). Initiation and maintenance of virus-induced gene silencing. *Plant Cell* 10: 937.

Saez-Rodriguez, J., Simeoni, L., Lindquist, J.A. et al. (2007). A logical model provides insights into T cell receptor signaling. *PLoS Comput Biol* 3 (8): e163.

Sander, J.D., Cade, L., Khayter, C. et al. (2011). Targeted gene disruption in somatic zebrafish cells using engineered TALENs. *Nat Biotechnol* 29: 697–698. doi:10.1038/nbt.1934.

Sanjana, N.E., Cong, L., Zhou, Y. et al. (2012). A transcription activator-like effector toolbox for genome engineering. *Nat Protoc* 7: 171–192. doi:10.1038/nprot.2011.431.

Sasaki, S., Yamagishi, N., and Yoshikawa, N. (2011). Efficient virus-induced gene silencing in apple, pear and Japanese pear using Apple latent spherical virus vectors. *Plant Methods* 7: 1.

Sather, D.N., Jovanovic, M., and Golenberg, E.M. (2010). Functional analysis of B and C class floral organ genes in spinach demonstrates their role in sexual dimorphism. *BMC Plant Biol* 10: 1.

Scofield, S.R., Huang, L., Brandt, A.S., and Gill, B.S. (2005). Development of a virus-induced gene-silencing system for hexaploid wheat and its use in functional analysis of the Lr21-mediated leaf rust resistance pathway. *Plant Physiol* 138: 2165–2173.

Seo, P.J., Hong, S.Y., Ryu, J.Y. et al. (2012). Targeted inactivation of transcription factors by overexpression of their truncated forms in plants. *Plant J Cell Mol Biol* 72 (1): 162–172.

Soula, H. and Chow, C.C. (2007). Stochastic dynamics of a finite-size spiking neural network. *Neural Comput* 19 (12): 3262–3292.

Takuhara, Y., Kobayashi, M., and Suzuki, S. (2011). Low-temperature-induced transcription factors in grapevine enhance cold tolerance in transgenic Arabidopsis plants. *J Plant Physiol* 168 (9): 967–975.

Tao, X. and Zhou, X. (2004). A modified viral satellite DNA that suppresses gene expression in plants. *Plant J* 38: 850–860.

Turnage, M.A., Muangsan, N., Peele, C.G., and Robertson, D. (2002). Geminivirus-based vectors for gene silencing in Arabidopsis. *Plant J* 30: 107–114.

Waterhouse, P.M., Wang, M., and Lough, T. (2001). Gene silencing as an adaptive defence against viruses. *Nature* 411: 834–842.

Weber, E., Gruetzner, R., Werner, S. et al. (2011). Assembly of designer TAL effectors by Golden Gate cloning. *PLoS One* 6: e19722. doi:10.1371/journal.pone.0019722.

Weng, L., Zhao, F., Li, R., and Xiao, H. (2015). Cross-talk modulation between ABA and ethylene by transcription factor SlZFP2 during fruit development and ripening in tomato. *Plant Signal Behav* 10 (12): e1107691.

Wesley, S.V., Helliwell, C.A., Smith, N.A. et al. (2001). Construct design for efficient, effective and high-throughput gene silencing in plants. *Plant J Cell Mol Biol* 27: 581–590.

Wolfe, S.A., Nekludova, L., and Pabo, C.O. (2000). DNA recognition by Cys2His2 zinc finger proteins. *Ann Rev Biophys Biomol Struct* 29: 183. doi.

Wright, D.A., Thibodeau-Beganny, S., Sander, J.D. et al. (2006). Standardized reagents and protocols for engineering zinc finger nucleases by modular assembly. *Nat Protoc* 1: 1637–1652. doi:10.1038/nprot.2006.259.

Xiao, A., Hu, Y.Y., Wang, W.Y. et al. (2011). *Yi chuan = Hereditas* 33 (7): 665–683.

Zamore, P.D., Tuschl, T., Sharp, P.A., and Bartel, D.P. (2000). RNAi: double-stranded RNA directs the ATP-dependent cleavage of mRNA at 21 to 23 nucleotide intervals. *Cell* 101: 25–33.

Zegzouti, H., Jones, B., Frasse, P. et al. (1999). Ethylene-regulated gene expression in tomato fruit: characterization of novel ethylene-responsive and ripening-related genes isolated by differential display. *Plant J* 18: 589–600.

Zhang, F., Cong, L., Lodato, S. et al. (2011). Efficient construction of sequence-specific TAL effectors for modulating mammalian transcription. *Nat Biotechnol* 29: 149–153. doi:10.1038/nbt.1775.

Zhang, Y., Ge, X., Yang, F. et al. (2014). Comparison of non-canonical PAMs for CRISPR/Cas9-mediated DNA cleavage in human cells. *Sci Rep* 4: 5405. doi:10.1038/srep05405.

Zhu, M., Chen, G., Zhou, S. et al. (2014). A new tomato NAC (NAM/ATAF1/2/CUC2) transcription factor, SlNAC4, functions as a positive regulator of fruit ripening and carotenoid accumulation. *Plant Cell Physiol* 55 (1): 119–135.

Zu, Y., Tong, X., Wang, Z. et al. (2013). TALEN-mediated precise genome modification by homologous recombination in zebrafish. *Nat Methods* 10 (4): 329–331.

3

MADS-box Transcription Factors Necessary for Flesh Fruit Development and Ripening

3.1 Introduction

MADS-box genes encode a large family of transcription factors that play important regulatory roles in eukaryotic development. The name "MADS" was derived from the four family members that were first discovered in different species and represents a combination of the first letters of their names: MCM1, AGAMOUS, DEFICIENS, and SRF. Mini chromosome maintenance1 (MCM1) is a regulator of yeast cell-type-specific genes. AGAMOUS and DEFICIENS are products of genes specific to floral organ determination in *Arabidopsis thaliana* and *Antirrhinum majus*, respectively (Sommer et al. 1990; Yanofsky et al. 1990). Serum Response Factor (SRF) is a human serum response factor (Messenguy and Dubois 2003). Sequence analysis has shown that all of these proteins contain a highly conserved motif consisting of 56–58 amino acids referred to as the MADS-box. MADS-box gene family members are transcription factors involved in many important biological processes in eukaryotes and are widely present in yeasts, plants, insects, nematodes, invertebrates, and mammals (Alvarez-Buylla et al. 2000; Becker and Theissen 2003; Nam et al. 2003).

3.2 MADS-box Gene Classification

In animals and fungi, the type I MADS-box genes include *ARG80/SRF-like* genes, while in plants, the type I genes are the M-type MADS-box genes (Figure 3.1). Type II MADS-box genes in animals and fungi include *MEF2-like* genes; the corresponding genes in plants are the MIKC-type MADS-box genes (Arora et al. 2007; Díaz-Riquelme et al. 2009). MIKC-type genes are typical MADS-box genes that are specific to plants. Based on the sequence divergence at the I-domain, they can be divided into two subgroups: $MIKC^{C}$-type and $MIKC^{*}$-type MADS-box

Transcriptional Regulation of Flesh Fruit Development and Ripening, First Edition. Yunbo Luo.

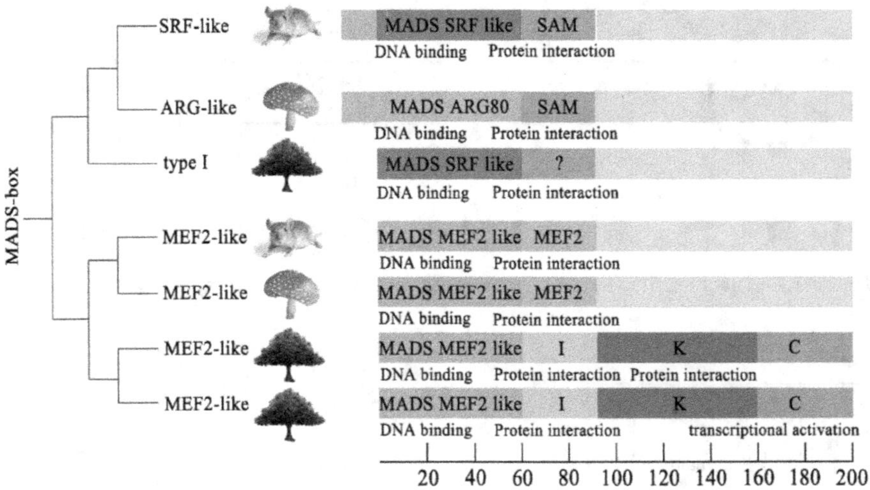

Figure 3.1 MADS-box protein classification (kontur-vid/Adobe Stock).

genes (Díaz-Riquelme et al. 2009). In *Arabidopsis thaliana*, MADS-box genes can be divided into five groups: Mα, Mβ, Mγ, Mδ, and MIKC-type genes (Parenicová et al. 2003). Mα, Mβ, and Mγ are type I MADS-box genes, whereas Mδ and MIKC belong to the MIKC*-type and MIKC^C-type, respectively, of the type II MADS-box genes. Studies have shown that type I MADS-box genes have undergone a higher rate of birth-and-death evolution than type II MADS-box genes. This difference may be because type I MADS-box genes have fewer important functions and are expressed at very low levels. As a result, type I MADS-box genes have fewer functional limitations and higher frequencies of gene duplications. Compared to type II MADS-box genes, type I MADS-box genes experience weaker purifying selection (De et al. 2003; Nam et al. 2004).

3.3 Motifs of the MADS-box Genes

The most intensively studied MADS-box genes are the MIKC-type MADS-box genes in plants; the structures and functions of these genes have been studied in depth (Figure 3.2). In addition to the conserved MADS-box domain, the proteins encoded by these genes also contain a semiconserved keratin-like (K-) domain, a less conserved intervening (I-) domain, and a C-terminal (C-) domain (Kaufmann et al. 2005).

The MADS-domain is the primary site of DNA binding. It is a sequence-specific DNA-binding domain; it can form dimers or higher structural complexes that recognize and bind the specific CArG box on the promoter, as shown in Figure 3.3 (Pollock 1990; de Folter and Angenent 2006; Ito et al. 2008; Smaczniak et al. 2017).

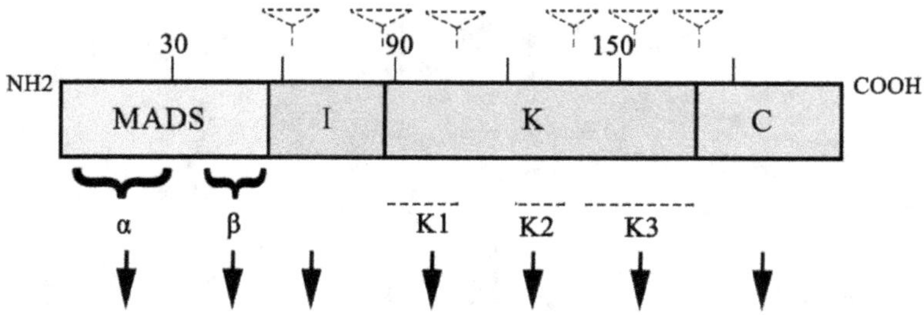

Figure 3.2 Diagram showing MIKC-type MADS-box protein structure.

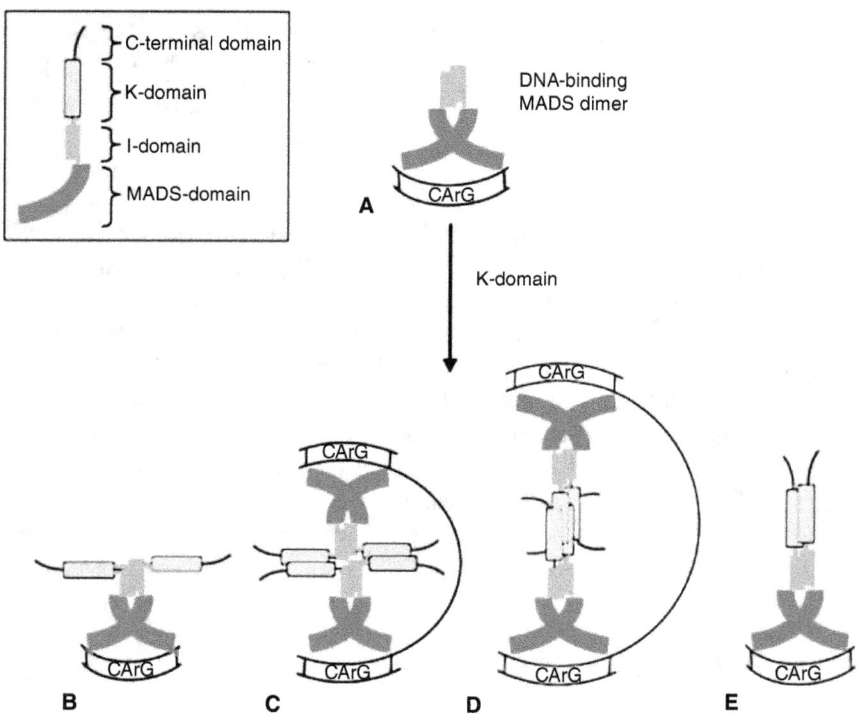

Figure 3.3 Model of the interactions among MIKC-type transcription factors.

MADS-box protein dimers consist of three structural units. The first unit is the DNA-binding region. Two amphipathic α-helices of a MADS protein dimer fold into an antiparallel coiled-coil parallel to the minor groove of the CArG box recognized by the MADS protein. The protein is also in contact with the DNA phosphate backbone and surrounds the DNA molecule through the coiled-coil, while the chain extending from the helical N-terminus reaches over the DNA backbone, penetrates the minor groove of the CArG box in the DNA double helix, and binds to the DNA molecule. The second structural unit is a four-stranded, antiparallel β-sheet formed by two β-strands from each monomer (βI and βII). It is oriented parallel to the coiled-coil structure formed by the α-helices. The third structural unit is located on top of the β-sheet and is formed by an irregular coiled-coil and a short α-helix (Messenguy and Dubois 2003).

The I-domain shows great variation in sequence and length (27–42 amino acids); it helps the transcription factor dimer form a complex with the DNA.

The semiconserved K-domain is a 70-aa region. Its name was derived from its homology to the coiled-coil region of keratin. The characteristics of the K-domain are that it is composed of hydrophobic amino acid residues, and the spanning allows the formation of amphipathic α-helices. It can be divided into three subdomains: K1, K2, and K3. The K-domain is required for the formation of MADS-box protein dimers.

The C-domain is the least conserved domain, and it primarily includes hydrophobic amino acid residues. Studies have shown that the C-domain is involved in the transactivation of higher-order MADS protein complexes and of the activation of polymers (Mueller and Nordheim 1991; Lai et al. 2019).

3.4 Functional Form of MADS-box Proteins

Studies have found that the dimerization of plant MADS-box transcription factors is an important means of regulating the transcription of their downstream genes. Immunoprecipitation experiments showed that *Arabidopsis* AG, APETALA1 (AP1), and AGAMOUS-like 1 (AGL1) can form dimers (Figure 3.4). MADS-box transcription factors bind as dimers to the CArG box on the promoter to regulate the expression of target genes, thereby regulating developmental processes.

3.5 Functions of the MADS-box Family

MADS-box transcription factors are involved in many aspects of plant development, including regulation of flowering time, reproduction, organs, meristems, and root structure (Kieffer and Davies 2001; Causier et al. 2002), as shown in Figure 3.5. Most studies of their functions have concentrated on the flower.

3.5.1 The Role of MADS-box Genes in Flower Development

Research in recent decades has shown that the MADS transcription factor family plays a crucial role in flower development (Coen and Meyerowitz 1991; de Folter and Angenent 2006). In 1991, Coen and Meyerowitz proposed the ABC model of

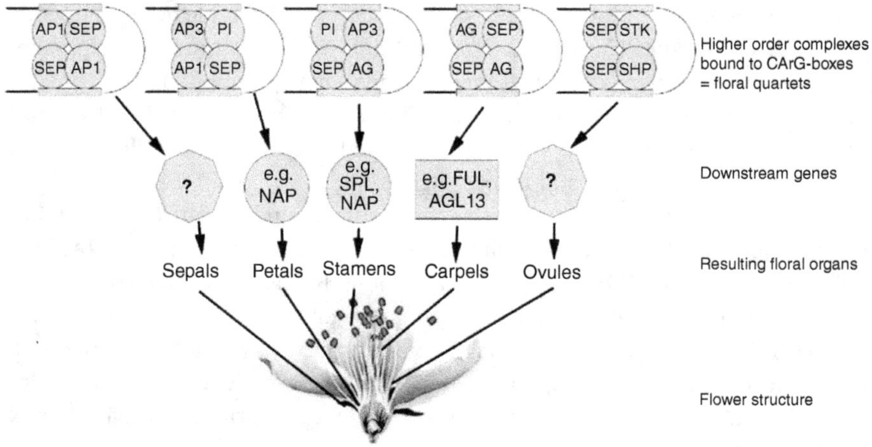

Figure 3.4 Regulation of floral development in *Arabidopsis thaliana*.

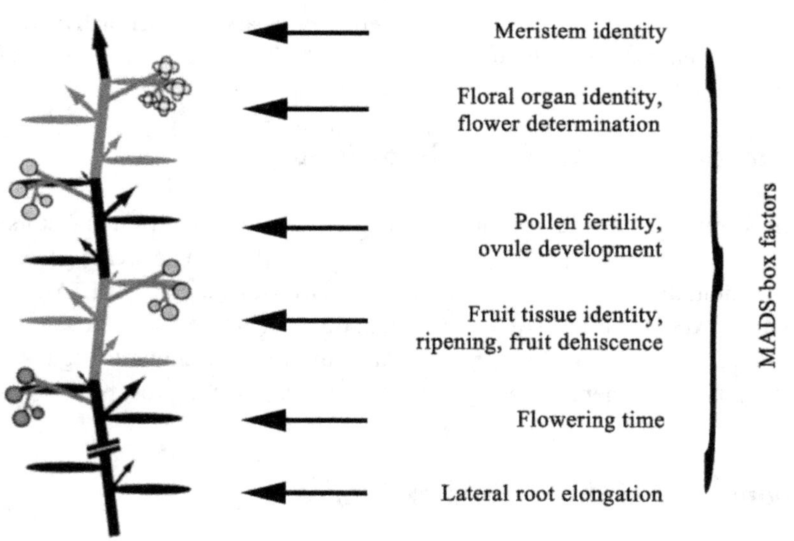

Figure 3.5 MADS-box transcription factors are involved in many developmental processes in plants.

floral organ development, which has been widely accepted (Coen and Meyerowitz 1991). Genes controlling the characteristics of floral organs can be divided into three classes, A, B, and C, based on their functions. "A" genes control the development of floral organs in whorls 1 and 2; loss of their function can cause the development of a carpel instead of a sepal in whorl 1 and a stamen instead of a petal in whorl 2. “B” genes control the development of floral organs in whorls 2 and 3; loss of their functions can lead to the development of a sepal instead of a petal in whorl 2 and a carpel instead of a stamen in whorl 3. “C” genes control the floral organ development of whorls 3 and 4; loss of their function can lead to the development of a petal instead of a stamen in whorl 3 and a sepal instead of a carpel in whorl 4. Most members of classes A, B, and C are MADS transcription factors. MADS-box genes also act at various stages during flower development to determine the characteristics of floral primordium and organs (Kaufmann et al. 2005).

Class A genes include *Arabidopsis APETALA1* (*AP1*) and *APETALA2* (*AP2*). However, *AP2* is not a MADS-box gene. It belongs to a small gene family that is specific to plants and encodes a transcription factor. In addition to regulating floral organ identity, AP1 can also regulate floral meristem identity during the early stages of flower development. AP1 transcripts can be detected in the floral primordium, and AP1 is responsible for the development of sepals and petals during subsequent flower development.

Class B genes include *Arabidopsis AP3* and *PI*, *Antirrhinum majus DEFICIENS* (*DEF*) and *GLOBOSA* (*GLO*), and *pMADS1*, *pMADS1*/*FBP3* and *FBP1* in petunia. The two class B genes in *Arabidopsis thaliana*, *APETALA3* (*AP3*) and *PISTILLATA* (*PI*), are responsible for regulating petal and stamen identity; *ap3pi* double mutants show sepals and carpels mutated from petals and stamens. During their regulation, AP3 and PI expression requires the activation of at least three genes: AP1, LFY, and UNUSUAL FLORAL ORGANS (*UFO*). Plants with the strong *lfy* allele had no stamen, while plants with the weak *lfy* allele had abnormal stamens. Studies have shown that *AP3* and *PI* expression is reduced in *lfy* and *ufo* mutants, indicating that AP3 and PI activities are under the regulation of LFY and UFO and that *AP3* expression requires both LFY and UFO. AP1 and LFY exhibit functional redundancy in activating B genes. AP1 regulates UFO activity; *UFO* transcripts cannot be detected in the sepals of *ap1* mutants. In addition, petal development is inhibited in *ap1* mutants. The regulation of petal identity by *AP1* (an A gene) is achieved through its regulation of UFO. Together, AP1 and UFO regulate the expression of *AP3* and *PI* genes to control petal identity (Parenicová et al. 2003). Suppression of B (*TrihDEFa-SRDX*) function may reduce the identity of petaloid tepals in whorls 1 and 2 of transgenic *Lilium sp.* (Otani et al. 2020). Furthermore, there are other genes that negatively regulate B genes, including *Arabidopsis SUPERMAN* (*SUP*, also called *FLO10*). *Sup* mutants have extra stamens at the location of carpels. *AP3* and *PI* are detected in whorl 4, indicating that SUP can inhibit *AP3* and *PI* expression in whorl 4. *SUP* expression in whorl 3 is dependent on AP3 and PI activity, as shown by the fact that in *ap3pi* mutants, decreased *SUP* expression is observed at early stages, and no *SUP* expression occurs in later stages. This finding suggests that AP3 and PI can activate *SUP* expression and that *SUP* expression can then inhibit *AP3* and *PI* expression (Ma 2005).

Class C genes include *AGAMOUS* (*AG*) in *Arabidopsis thaliana*, *PLENA* in *Antirrhinum majus*, and *pMADS3* in petunia (Kater et al. 2006; Morel et al. 2018; Ito et al. 2019). C genes regulate stamen and carpel identity, and their expression is also controlled by many other genes. LFY directly and positively regulates *AG*, a class C gene in *Arabidopsis thaliana*. Plants carrying the strong *lfy* allele show missing petals and stamens and have irregularly fused carpels. WUSCHEL (WUS) can also positively regulate *AG*. In the floral meristem, *WUS* is expressed before *AG* is expressed; *wus* mutants exhibit decreased flower number, loss of carpels, and numerous stamens in some flowers. Reduced *WUS* expression leads to a decrease in *AG*-expressing regions, while the ectopic expression of *WUS* can cause ectopic growth of stamens and carpels. Some studies concluded that both the *WUS* and *LFY* genes regulate *AG* expression by binding to the cis-regulatory sequences of *AG*. The absence of stamens and carpels in *ag* mutant flowers shows that *AG* is necessary for plant reproduction. Although *AG* is a gene that regulates carpel development, an *AG*-independent carpel developmental pathway also exists in plants. In this pathway, *SHATTERPROOF* (*SHP*), a D gene, is a major player. In *ap2* mutants, *AG* is expressed in the entire flower; sepals are converted to carpelloid organs in the first-whorl organs. However, a similar phenomenon in which carpelloid organs replace sepals is also observed in the first-whorl organs of the *ap2ag* double mutant, indicating the existence of another gene that regulates carpel development independently of AG. Studies have shown that *SHP* is expressed in ectopic carpels of *ap2* mutants. However, in *ap2agshp1shp2* quadruple mutants, conversion of sepals to carpelloid organs in the first-whorl organs was not observed. This result suggests that SHP can control the AG-independent development of carpels and ovules in *ap2ag* mutants. Under normal conditions, the regulation of carpel development by SHP is dependent on AG activity. However, in *ap2* mutants, *SHP* expression in the first-whorl organs is independent of AG (Pinyopich et al. 2003). Morel et al. (2018) demonstrated that AG and PLENA proteins are essential for nectary development in petunia, while their overexpression is sufficient to induce ectopic nectaries on sepals. Sugiyama et al. (2019) readdress the function of class C genes in rice (*Oryza sativa*) and provide evidence to suggest that, in contrast to Arabidopsis, rice class C genes are not key regulators of carpel specification.

Studies conducted in recent years have found that in addition to the A, B, and C genes, the *SEPALLATA* (*SEP*) genes also regulate floral organ identity (Pelaz et al. 2000; Zhang et al. 2022). Angenent et al. (1995) first discovered such a function of *SEP* in a cosuppression experiment using petunia and tomato. They found that *SEP* could regulate the development of petals, stamens, and carpels. *Arabidopsis* single and double mutants of *SEP1* (*AGL2*), *SEP2* (*AGL4*), and *SEP3* (*AGL9*) did not exhibit significant changes in floral organ identity, but the triple mutants produced flowers in which all organs developed as sepals. This phenotype is similar to that of the *bc* double mutants; therefore, it is believed that the regulation of petal, stamen, and carpel identity by B and C genes requires *SEP1/2/3* (Pelaz et al. 2000). Molecular and genetic experiments have shown that ectopic *SEP3* expression ectopically activates *AP3* and *AG*, indicating that SEP3 can activate B and C genes (Castillejo et al. 2005). Ditta et al. (2004) noted that *AGL3* is functionally redundant to *SEP1*, *SEP2*, and *SEP3* in regulating sepal and meristem identity. Therefore, they named *AGL3* as *SEP4*. Similarly, single mutants of *SEP4* did not exhibit any morphological changes in floral organs, but in *sep1sep2sep3sep4* quadruple mutants, the floral organs were converted to leaflike organs. Based on molecular genetic studies and studies of protein–protein interactions, an ABCE model has been proposed. In this model, A genes and *SEP1/2/3/4* (E genes) regulate sepals, whereas A and B genes, as well as *SEP1/2/3/4* genes, regulate petals. B, C, and *SEP1/2/3/4* genes regulate stamens, and C genes and *SEP1/2/3/4* genes regulate carpels (Ma 2005).

3.5.1.1 Control of Flowering Time

Flowering marks the beginning of the reproductive stage. Many MADS-box genes play roles in the conversion from plant vegetative growth to reproductive growth. For example, *AGL24* and *SCO1* promote flowering, while *FLC* and *SVP* inhibit flowering (Michaels et al. 2003a). In *Arabidopsis thaliana*, three pathways control flowering time. The first of these is the autonomous pathway. In this regulatory pathway, Flowering Locus C (FLC) plays a major role by negatively regulating the transition from vegetative growth to reproductive growth. Michaels et al. (2003b) showed that *flc* mutants exhibited an early flowering phenotype, indicating that FLC can inhibit flowering. The second pathway is the photoperiod pathway, in which long daylight induces flowering. The *CONSTANS* (*CO*) gene, which directly targets *SUPPRESSOR OF OVEREXPRESSION OF CONSTANS1* (*SOC1*), a MADS-box family member, plays major roles in this regulatory pathway. Mutation in *CO* leads to delayed flowering. The third pathway is the photoperiod-independent pathway that is associated with the plant hormone gibberellic acid (GA). It controls flowering by activating *LFY*. MADS-box genes, especially *SOC1*, are involved in all three of the regulatory pathways that regulate flowering. Thus, MADS-box genes are important factors that regulate flowering time.

3.5.1.2 Regulation of Ovule Development

Ovules are organs that are specific to seed plants; they develop into seeds after fertilization. The study of ovule development has important theoretical and practical significance. Currently, ovule development is mostly studied in petunia and *Arabidopsis thaliana*, and some genes related to ovule development have been discovered. Many of these genes are MADS-box genes.

1) The functions of FBP7 and FBP11 in ovule development

In addition to identification of the A, B, C, and D genes, other D genes that determine ovule development have been discovered. To date, there have been few studies of the functions of D genes. The *FLORAL BINDING PROTEIN 7* (*FBP7*) and *FBP11* genes in petunia are the most extensively studied D genes. They were isolated from petunia mutants in 1995 by Angenent et al., who found that mutation of these two genes caused spaghetti-shaped structures to develop in the positions normally occupied by ovules. It is believed that these two MADS-box genes are a new class of MADS-box genes and that their functions are related to ovule development. Colombo et al. (1995) generated transgenic petunia overexpressing *FBP11* and observed ovule-like structures on the adaxial side of the sepals and the abaxial side of the petals. The expression pattern of *FBP11* was similar to that of *FBP7* and the C genes *FBP6* and *pMADS3*. *FBP11* is considered an ovule-specific gene. Immink et al. (2002) also found that FBP11 could form heterodimers with FBP2, FBP5, and FBP9 (encoded by E genes) and regulate ovule development in petunia.

2) The function of STK genes in ovule development

Benfey et al. (1990) overexpressed *SEEDSTICK* (*STK*) under the CaMV35S promoter in transgenic *Arabidopsis thaliana* and found that it is necessary for induction of the formation of abnormal ovules. Pinyopich et al. (2003) found that the *STK* gene had high similarity to *FBP7* and *FBP11* in petunia and that *STK*, *SHP1*, and *SHP2* played important roles in ovule development. Favaro et al. (2003) found that in transgenic plants overexpressing *STK*, sepals and leaves were transformed to capelloid organs, while ovules developed from the capelloid organs. Pinyopich et al. (2003) found that *ag* mutants overexpressing *STK* could produce ovules at the locations of sepals and petals, possibly due to the induction *SHP1* and *SHP2* expression by STK. The functions of the *SHP* genes in ovule development have been studied. Liljegren et al. (2000) found that, similar to *STK* overexpression, *SHP1* and *SHP2* overexpression led to the formation of ovules at the location of sepals. Pinyopich et al. (2003) found that *SHP* was induced by AG. Because *SHP* could not be induced in *ag* mutants, carpel formation was completely inhibited in the mutants. However, in *agap2* mutants, *SHP1*, *SHP2*, and *STK* were found to be expressed in the first-whorl organs, indicating that AP2 negatively regulates these two *SHP* genes either directly or indirectly. Pinyopich et al. (2003) also found that no ovules formed in *shp1shp2* mutants. Moreover, *SHP1* and *SHP2* were functionally redundant to *STK* in promoting ovule identity. In *stkshp1shp2* triple mutants, normal ovule and seed development was completely disrupted. Some ovules were converted to leaf- or carpel-like structures. Nevertheless, missense mutations or reduced expression of *STK* orthologues give rise to seedless phenotypes in cultivated grape (Ocarez and Mejía 2016; Royo et al. 2018). In addition, in the presence of SEP, STK can interact with SHP1, SHP2, and AG, which promote ovule development.

3) The functions of SEP genes in ovule development

Yeast two-hybrid experiments revealed that SEP3 protein interacts with proteins that promote carpel and ovule identity and that SEP3 is related to ovule development (Kitazawa et al. 2022). Favaro et al. (2003) found that a reduction in SEP activity could lead to phenotypic changes in the ovule similar to those that occur in *stkshp1shp2* triple mutants. Normal ovule and seed development was disrupted; some ovules were converted to leaf- or carpel-like structures. These results show that *SEP* genes are required for normal ovule development. Furthermore, although SEP3 alone can promote normal ovule development, it only functions when SEP2 activity is present. One active *SEP2* allele was not sufficient to promote ovule development, although one *SEP3* allele could support normal ovule development. Thus, these two *SEP* genes are not completely redundant. Moreover, there are distinct differences in the expression patterns of *SEP2* and *SEP3*.

Beyond the discovery of MADS-box genes that affect ovule development in model plant species, homologous genes have also been found in other species. Lopez-Dee et al. (1999) isolated Os*MADS13* from rice and identified its functions in ovule and seed development in rice. Xu et al. (2004) isolated the Ho*MADS1* gene in *Hyacinthus* and found that Ho*MADS1* overexpression produced carpelloid organs, including ovules. Based on this, it was concluded that Ho*MADS1* plays important roles in determining carpel and ovule identity. Tzeng et al. (2003) isolated two MADS-box genes, L*MADS2* and Eg*MADS1*, from lily. Analysis showed that they were highly homologous to *FBP7* and *FBP11*. L*MADS2* is mostly expressed in ovules, with weak expression in carpels, while Eg*MADS1* is specifically expressed in ovules. Zhang et al. (2022) isolated the *VvMADS39*, a class E MADS-box gene orthologous to *Arabidopsis SEP2*, from grapevine. *VvMADS39* plays an important role in maintaining floral meristem characteristics, fruit and ovule development.

Other studies have shown that the MADS-box gene family is also involved in regulating various important physiological processes, including regulation of seed formation, vegetative organ development, fruit ripening, flower and fruit abscission, root development, and seed embryonic development. Thus, MADS-box genes play various roles in plant development, regulating different aspects of development or acting at different levels (Causier et al. 2002; Zhang et al. 2022).

Based on the technology of sequencing, molecular biology and bioinformatics analysis, numerous members of the MADS-box gene family and their functions have been revealed. However, the results so far have revealed just the tip of the iceberg of the MADS-box gene family. The knowledge of the MADS-box gene family in floral organogenesis in plant needs to be further revealed.

3.5.2 The Regulation of Fruit Ripening by MADS-box Transcription Factors

As a model material for studying the development, ripening, and senescence of fleshy fruits, the tomato has been studied intensively and systematically worldwide.

The development of tomato fruits begins at fruit setting. Early fruit development mainly involves cell division and differentiation. At approximately 14 days after flowering, cell division ceases, and the plant enters a rapid cell enlargement stage during which the fruit volume rapidly increases. After entering the mature green stage, the fruit volume no longer changes. The transition from the mature green stage to the breaker stage is an important stage of tomato fruit ripening. During this stage, numerous physiological and biochemical reactions are activated. The most important of these is System II ethylene biosynthesis and the occurrence of the climacteric peak. The plant hormone ethylene is the main factor that activates and promotes tomato fruit ripening. As a large quantity of ethylene is released, enzymes related to cell wall degradation, including polygalacturonase (PG), pectinesterase (PME), and β-galactosidase, begin to appear in the fruit (Giovannoni 2004). The fruits begin to soften, accompanied by changes in fruit color that are used as a reference for identifying fruit ripening under the naked eye. The change from green to red color reflects the process of chlorophyll degradation and the accumulation of carotenoids, mainly lycopene and β-carotene.

3.5.2.1 The Effect of Tomato RIN on Fruit Ripening

Among the MADS-box transcription factors related to the regulation of fruit ripening in tomato, Ripening-Inhibitor (RIN) is the most extensively studied. RIN belongs to the MIKC subfamily of the MADS family; most studies related to RIN are based on *rin* mutants. The tomato *ripening-inhibitor* (*rin*) mutant has long been in the research spotlight. Its fruits fail to ripen normally; ripening is severely delayed, and there is a lack of ethylene and of a climacteric peak. There is no accumulation of carotenoids, and the degree of fruit softening is extremely low. In addition, tomato *rin* mutants exhibit enlarged sepals and altered inflorescence structure (Causier et al. 2002; Vrebalov 2002).

In tomato *rin* mutants, positional cloning of the *rin* locus revealed two tandem MADS-box genes, Le*MADS-RIN* and Le*MADS-MC*. Le*MADS-RIN* is mostly expressed in fruits, and its expression increases as the fruit ripens. Le*MADS-MC* is mostly expressed in sepals, petals, and carpels. Results obtained using transgenic sense and antisense *rin* mutants have shown that MADS-RIN (RIN) affects fruit ripening while MADS-MC (MC) regulates sepal development and inflorescence determinacy (Vrebalov 2002). In wild-type AC (Ailsa Craig) tomatoes, the two tandem *MADS-RIN* and *MADS-MC* genes, which are separated by 2.6 kb, encode proteins containing 242 and 244 amino acid residues, respectively (Vrebalov 2002). In tomato *rin* mutants, partial deletion of the 3′-end of *RIN* and the 5′-end of *MC* caused the two genes to fuse, forming a *RIN-MC* gene (Figure 3.7). The fusion *RIN-MC* transcripts could be detected by amplification (Vrebalov 2002). The fusion of the *RIN* and *MC* gene resulted in failure of the *RIN* gene to be normally transcribed and consequent failure of RIN protein expression, thereby affecting the regulation of tomato fruit ripening by *RIN*. Anyway, we have shown that *rin* fruit in fact contain abundant RIN-MC mRNA, which is translated. The encoded protein is located in the nucleus, interacts with other MADS-box transcription factors. Overexpression of RIN-MC in transgenic wild-type cv Ailsa Craig tomato fruits impaired several ripening processes (Figure 3.8), and downregulating *RIN-MC* expression in the rin mutant was found to stimulate the normal yellow mutant fruit to produce a weak red color, suggesting a distinct negative role for RIN-MC in tomato fruit ripening (Figure 3.9). Tomato RIN-MC fusion plays a negative role in ripening and

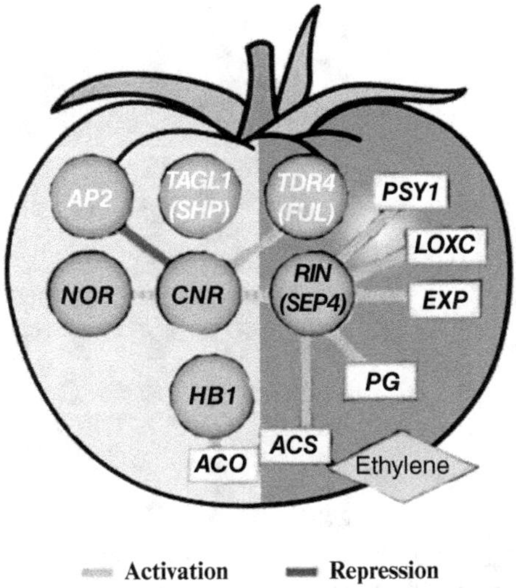

Figure 3.6 Main regulators of tomato fruit ripening. The blue circles represent transcription factors.

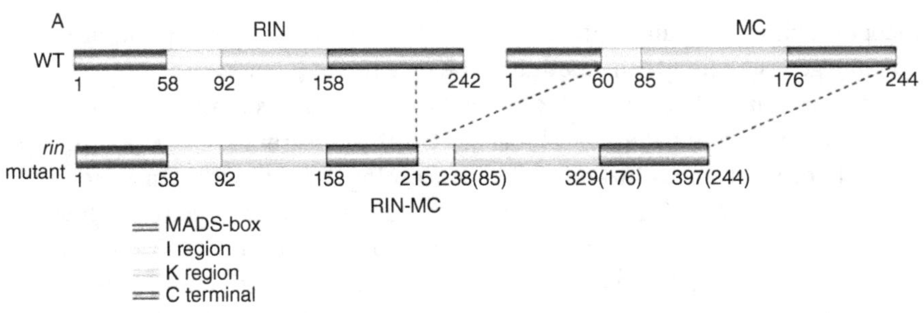

Figure 3.7 Mechanism of RIN-MC fusion in tomato *rin* mutants (Li et al. 2018/Oxford University Press).

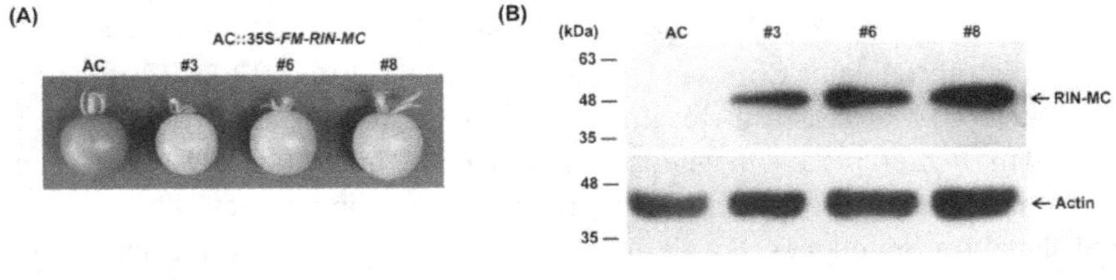

Figure 3.8 Effects on tomato ripening of the overexpression of RIN-MC in wild-type (AC) fruits. Results from three transgenic lines (3, 6, and 8) are compared with the wild-type control. A, Phenotypes. B, Levels of RIN-MC protein. (Li et al. 2018/with permission from Oxford University Press).

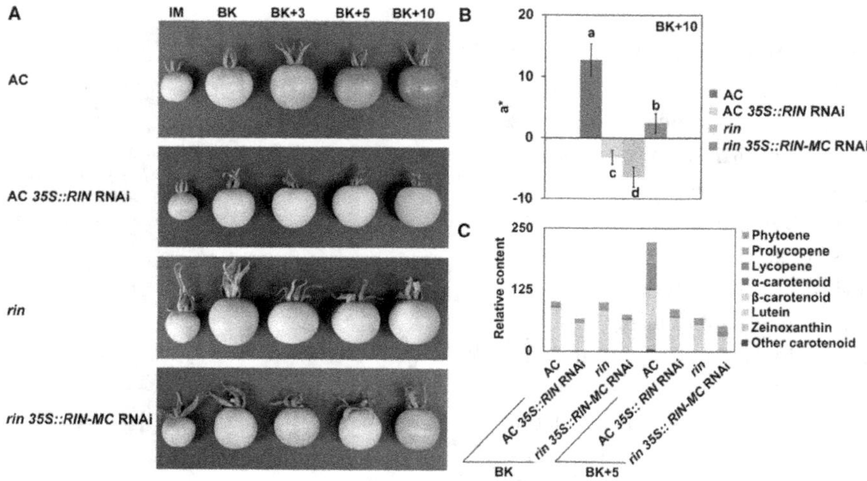

Figure 3.9 Effects of RIN and RIN-MC on fruit color and the accumulation of carotenoids. A, Phenotypes of four different lines: the wild type (AC), the wild type expressing a RIN-silencing construct, the rin mutant, and the rin mutant expressing a RIN-MC-silencing construct. B and C, Color (a*; B) and relative contents of carotenoids (C) (Li et al. 2018/with permission from Oxford University Press).

encodes a chimeric transcription factor that modulates the expression of many ripening genes, thereby contributing to the rin mutant phenotype.

In tomato *rin* mutants, loss of *RIN* function causes a severe ripening-inhibited phenotype, confirming the critical role of the RIN transcription factor in fruit ripening. Recent advances in knowledge concerning the transcription factor RIN that regulate climacteric fruits ripening indicated discrepancies between the phenotypes of natural mutant *rin* and CRISPR/Cas9-induced mutations have called into question the precise roles of the regulatory gene in the ripening control network (Ito et al. 2017; Li et al. 2020). Anyway, RIN affects fruit ripening mainly by regulating the expression of downstream target genes, thereby affecting the entire fruit-ripening process. In recent years, through differential comparative transcriptomic and proteomic analyses, chromatin immunoprecipitation (ChIP) coupled with DNA microarray analysis (ChIP-chip) (Qin et al. 2012; Fujisawa et al. 2013), high-throughput ChIP-seq (Martel et al. 2011), and ChIP-PCR validation (Fujisawa et al. 2012), fruit-ripening-related genes affected by RIN have been studied in *rin* mutants and in wild-type AC tomatoes. The results have shown that RIN affects many processes during fruit ripening, including ethylene biosynthesis and signal transduction, cell wall metabolism, the formation and metabolism of carotenoids, and the formation of aromatic volatile compounds (Figure 3.6). RIN also regulates fruit ripening by affecting the transcription of other transcription factors related to fruit ripening such as *COLOURLESS NON-RIPENING* (*CNR*),

TDR4, and *HB-1* (Martel et al. 2011; Fujisawa et al. 2012; Qin et al. 2012). In addition, RIN can directly bind to the promoter regions of fruit-ripening-related long noncoding RNAs (lncRNAs) and further regulate its transcription, such as lncRNA2155 (Yu et al. 2019).

1) RIN regulates transcription factors related to fruit ripening

As an important regulator of tomato fruit ripening, RIN regulates other transcription factors related to ripening and thereby affects the fruit-ripening process. First, RIN has self-regulatory capability, and it can regulate other transcription factors such as NOR, HB-1, CNR, and TDR4 (Martel et al. 2011). No-ripening (NOR) is a member of the NAC family of transcription factors and plays important roles in fruit ripening. Like *rin* mutants, *nor* mutants exhibit ripening inhibition (Giovannoni 2007). HB-1 is a transcription factor belonging to the HD-zip family; it affects fruit ripening by positively regulating the expression of *ACC OXIDASE 1* (*ACO1*), which encodes a key enzyme in the ethylene biosynthetic pathway (Lin et al. 2008). CNR is an SBP family transcription factor; it is also a key regulatory factor for fruit ripening. Tomato *cnr* mutants exhibit a nonripening phenotype; the pericarp tissue fails to accumulate the pigment that is normally present in mature fruit (Eriksson et al. 2004; Manning et al. 2006). The transcription factor TDR4 is also a member of the MADS family and is involved in the regulation of fruit ripening. One study found that in *Vaccinium myrtillus* the homolog of TDR4 affects fruit ripening (Jaakola et al. 2010).

2) RIN regulates ethylene biosynthesis and metabolism during ripening

Tomato fruits are typical climacteric fruits, but in *rin* mutants, ethylene biosynthesis is severely inhibited, and the climacteric peak is absent. Thus, RIN may directly regulate the expression of key genes in the ethylene biosynthetic pathway and thereby affect the formation of ethylene during fruit ripening. During fruit ripening, ethylene biosynthesis is under the control of several key genes, including *ACS2*, *ACS4*, and *ACO1*. Additionally, mutations in the ethylene receptor gene *NEVER-RIPE* (*NR*/Le*ETR3*) can lead to ethylene insensitivity that inhibits fruit ripening. Moreover, E4 and E8 genes regulate the accumulation of ethylene during fruit ripening. Studies have shown that in *rin* mutants expression of the ethylene biosynthetic genes *ACS2*, *ACS4*, and *ACO1*, the ethylene receptor gene *NR*, and the ethylene accumulation-related genes *E4* and *E8* is significantly downregulated. Bioinformatics analysis of the RIN-binding site and ChIP-PCR verification showed that RIN could directly regulate the expression of *ACS2*, *ACS4*, *ACO1*, *NR*, *E4*, and *E8* (Fujisawa et al. 2013). Reduced amount of endogenous ethylene generated by *RIN-deficient* fruits is sufficient to initiate ripening in the absence by RIN, but ethylene treatment of RIN-deficient lines did not restore full ripening (Li et al. 2020), which indicated the requirement of both RIN and ethylene for full expression of ripening genes.

3) RIN regulates cell wall metabolism during fruit ripening

Cell wall formation and metabolism during fruit ripening are under the regulation of two key genes, *PG2a* and *EXP 1*. *PG2a* is a key enzyme involved in pectin degradation and is closely associated with fruit ripening. *Exp 1* affects fruit softening and is related to cell wall metabolism. It has been found that *PG2a* and *EXP 1* expression is severely downregulated in *rin* mutants, indicating that RIN directly regulates *PG2a* and *EXP 1* (Martel et al. 2011). In addition, the expression of *CEL2* was almost complete absence in the fruit of *RIN-deficient* mutants, and the decreased levels of *EXP 1*, *PL*, *XYL1*, *Mside1*, *Mside7*, *MAN1*, *MAN4a*, *TBG4*, *PG*, *PME1.9*, and *PME2.1* transcripts were also measured in mutant fruit (Li et al. 2020). Moreover, the levels of *XTH5* and *XTH8* transcripts were significantly upregulated in *RIN-deficient* mutant fruit, which might cause cell wall structural changes affecting texture or softening of RIN-deficient fruits during ripening (Li et al. 2020).

4) RIN regulates pigment accumulation in fruit ripening

Fruit ripening is accompanied by the accumulation of pigments, among which the carotenoids are the most important. During fruit ripening, the pigment biosynthesis-related gene *PHYTOENE SYNTHASE1* (*PSY1*) plays an important regulatory role in the formation of carotenoids (Fraser et al. 2007). *PSY1* encodes a fruit-specific isoenzyme in the carotenoid biosynthetic pathway. PSY1 is the rate-limiting enzyme in the biosynthesis of fruit pigment; it catalyzes the formation of phytoene. RIN, as the key regulatory factor for fruit ripening, can directly regulate the transcription of *PSY1*.

5) RIN regulates the formation of aromatic volatile compounds in the LOX pathway of fruit ripening

The formation of aromatic volatile compounds during fruit ripening is a key factor that affects fruit appearance and quality. It is estimated that over 400 aromatic volatiles are formed during tomato fruit ripening, but only a few of these, including hexanal, hexenal, hexenol, cis-3-hexen-1-ol, and 3-methylbutanal, determine the flavor of the tomato fruit. The formation of these aromatic volatile compounds occurs via multiple pathways, including the lipoxygenase (LOX) pathway of polyunsaturated fatty acids, the amino acid pathway involving the deamination and decarboxylation of amino acids, and the carotenoid pathway involving the oxidative cleavage of carotenoids. Among these pathways, the LOX pathway is one major pathway for the formation of aromatic volatile compounds during fruit ripening. There are at least three rate-limiting enzymes in the LOX pathway: LOX, hydroperoxide lyase (HPL), and alcohol dehydrogenase (ADH) (Schwab et al. 2008). LOX catalyzes the regio- and enantio-selective dioxygenation of unsaturated fatty acids (e.g., linoleic and α-linolenic acid). The products can be used by HPL to form aldehydes (hexanal and cis-3-hexenal), which can be further converted to hexenol and cis-3-hexen-1-ol by ADH. One study showed that the protein levels of LOX and ADH2 are significantly decreased in tomato *rin* mutants (Qin et al. 2012). Five genes containing the CArG box-binding site of RIN were also identified. ChIP and EMSA experiments confirmed that RIN could directly regulate *TomloxC* and *ADH2* expression. In another study, the decrease of four C5 and C6 volatiles associated in LOX pathway were measured in RIN-deficient fruits, which was related to the reduction in transcripts of biosynthetic pathways key genes *LoxC*, *LoxB*, *HPL*, and *ADH2* (Li et al. 2020).

6) Regulatory mode of RIN

Transcription factors regulate gene transcription by binding to specific sequences. Generally, this type of transcriptional regulation is not achieved through a single transcription factor but rather involves the formation of protein complexes consisting of one or more transcription factors. The protein complexes then specifically interact with target genes and regulate their expression. Many studies have shown that MADS-box proteins contain multiple domains related to protein–protein interactions and that they fulfill their biological functions through polymerization. Yeast two-hybrid technology was first used to study the protein–protein interactions of MADS-box proteins. Subsequently, Theißen and Saedler (2001) proposed the floral quartet model in which four MADS-box proteins first form two dimers; these dimers then bind to different sites on the same target gene and form a tetramer that regulates the transcription of the target gene. The tomato fruit-ripening-related transcription factor RIN usually forms homodimers, homotetramers, or heterodimers with FUL1 (TDR4) and FUL2 (MBP7) to fulfill its biological functions in fruit ripening. RIN can also form heterotetramers with FUL1/FUL2 and TAGL1 to regulate fruit ripening (Seymour 2013). The prediction model of protein polymers is shown in Figure 3.10 (Shima et al. 2013; Fujisawa et al. 2014).

7) Modifications of RIN regulatory mode

A recent study found that RIN function is lost due to the partial deletion of genomic DNA sequences in tomato *rin* mutants (Vrebalov et al. 2002). Moreover, transcription of the fusion *RIN-MC* gene occurs at a higher level during tomato fruit ripening, and the RIN-MC fusion protein also possesses transcription factor activity (Ito et al. 2017; Li et al. 2018). RIN-MC is an inhibitor of fruit ripening; RIN is critical for progression of ripening but is not required for ripening initiation (Ito et al. 2017; Li et al. 2018). The use of RNAi or CRISPR/Cas9 technology can effectively silence or knockout expression of the *RIN* gene. In transgenic tomato with *RIN*-RNAi, the expression level of *RIN* was reduced to less than 5% of the normal amount, and RIN proteins were almost undetectable (Li et al. 2018). The fruits of homozygous *RIN* knockout mutants generated by CRISPR/Cas9 technology displayed no RIN protein accumulation (Ito et al. 2015, 2017). Therefore, the identification of RIN target genes through the comparison of differentially

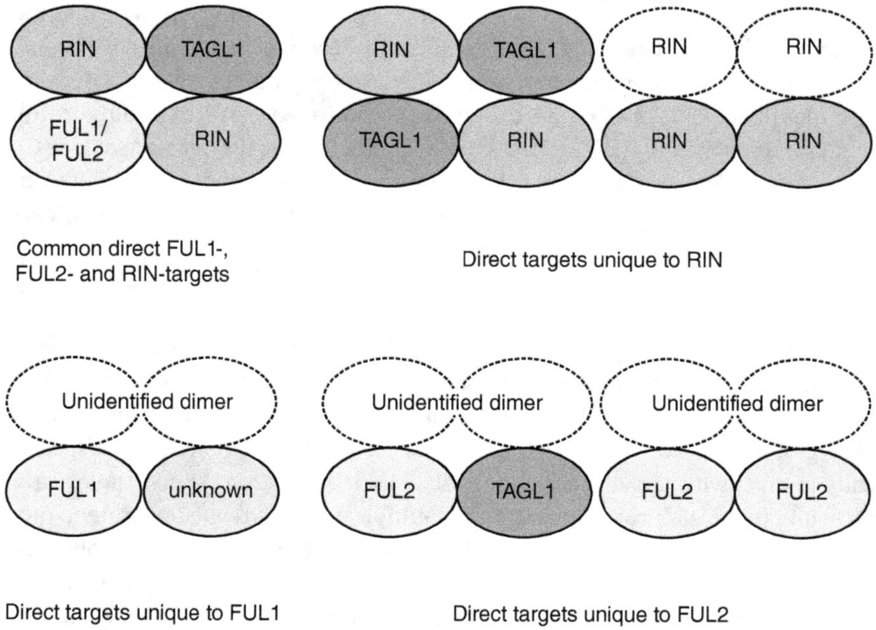

Figure 3.10 A prediction model of MADS-box protein complexes.

expressed genes between wild-type AC and *rin* mutant fruits needs to include a comparison of AC fruits with *RIN*-RNAi or *RIN*-CRISPR fruits (Li et al. 2019). When this comparison was taken into account, statistical analysis showed that RIN target genes contain 336 genes related to fruit ripening and quality; 115 genes were newly added, 150 genes were eliminated, and 19 genes were modified (analyzed from Li et al. 2018). Thus, using fruits generated by *RIN*-RNAi or *RIN*-CRISPR makes it possible to analyze the functions of RIN without solely relying on *rin* mutants. RIN-deficient fruits generated by CRISPR/Cas9 initiated partial ripening similarly to wild-type fruits, but never ripened completely, even supplied with exogenous ethylene (Li et al. 2020), which indicated that both RIN and ethylene and other factors are required for completion of the full fruit-ripening program.

3.5.2.2 The Effect of FRUITFULL on Tomato Fruit Ripening

The tomato transcription factors FUL1/TDR4 and FUL2/MBP7, as well as RIN, belong to the MADS-box family of transcription factors. The patterns of *FUL1* and *FUL2* expression are associated with fruit ripening. Studies have shown that the FUL1 and FUL2 proteins interact with RIN by forming heterotetramers (Leseberg et al. 2008; Martel et al. 2011; Shima et al. 2013).

1) Functional comparison between tomato FUL1 and FUL2

Nucleic acid sequence analysis of *FUL1* and *FUL2* showed that they differ only in their C-domains and at the 3′ end. Silencing only *FUL1* or *FUL2* did not result in a distinct phenotype in transgenic plants, although *FUL1* RNAi lines showed slight changes in fruit color. However, simultaneous interference with *FUL1* and *FUL2* in transgenic plants significantly inhibited the fruit-ripening process in tomato. Similar results were obtained in FUL1 and FUL2 knockout mutants generated by CRISPR/Cas9 technology as well as double mutants (Wang et al. 2019). The fruit color was orange, and ethylene emission was dramatically decreased compared to the wild type, indicating functional redundancy of *FUL1* and *FUL2* in the regulation of fruit ripening. Transcriptomic and metabolomic analyses have shown that tomato FUL1 and FUL2 are also involved in cell wall modification, the production of cuticle components and volatile compounds, and the accumulation of glutamic acid (Bemer et al. 2012). Transgenic plants overexpressing *FUL2* exhibit a series of abnormal phenotypes including thinner stalk, thinner pericarp, and changes in the outer pericarp and in fruit shape. However, no abnormal phenotypes were observed in transgenic plants overexpressing *FUL1*. Previous studies of *Arabidopsis ful* mutants showed that cell differentiation and expansion in the *ful* mutant silique are different from wild type; the cells are rounder and possess fewer cell layers. Based on the phenotype of transgenic tomato plants overexpressing *FUL2*, the FRUITFULL protein is predicted to participate in a conserved regulatory mechanism that controls cell expansion in the fruit of different plant species. In contrast, *FUL1* overexpression did not result in an abnormal phenotype, indicating that *FUL1* may have lost certain functions during evolution. From the differences in amino acid sequence between FUL1 and FUL2, it is speculated that the loss of this function in FUL1 may be related to the differences in the protein structures of FUL1 and FUL2. Leseberg et al. analyzed the interactions among 16 MADS-box proteins in tomato using the yeast two-hybrid method. They found that FUL2-interacting proteins included RIN, TM3, JOINTLESS, MBP18, MBP24, MBP21, MADS1, TAG1, and TM5, whereas FUL1 could only interact with TM3, JOINTLESS, and RIN. Moreover, the K-domain of MADS-box proteins was found to be closely associated with specific binding to the interacting proteins (Leseberg et al. 2008). The amino acid sequences of the K-domains in FUL1 and FUL2 share 84% homology, lower than the homology between the MADS-domain and the I-domain; therefore, the difference in the K-domain sequences could be the main reason for their functional differences. Additionally, a previous study showed that the C-domain of MADS-box proteins is not highly conserved and that differences in the C-domain are closely related to the differentiation of MADS-box protein function (de Folter and Angenent 2006). The amino acid sequences of the C-domains in FUL1 and FUL2 have 48% homology, suggesting that the difference in the C-domain could be a main cause of the differences in their functions.

2) The regulatory mode of FUL1 and FUL2 in fruit ripening

Through a comparative transcriptomic analysis of fruits from a FUL1/FUL2 silenced line and the wild-type line, FUL1 and FUL2 were found to regulate many genes involved in many processes during ripening, including *ACO1*, *ACS2*, and *ACS4* in the ethylene pathway. FUL1 and FUL2 also regulate the accumulation of carotenoids in tomato fruits; silencing both *FUL1* and *FUL2* resulted in tomato fruits not turning red but instead maintaining an orange color. In the fruit of a line in which both *FUL1* and *FUL2* were silenced, the expression of the key phytoene-biosynthetic gene *ZDS* and the key lycopene-biosynthetic gene *PSY1* was significantly inhibited. Thus, FUL1/FUL2 may act upstream of the carotenoid pathway.

To further clarify the binding sites of FUL1 and FUL2, a ChIP-chip analysis was conducted. The results showed that 50% of the FUL1-bound region and 46% of the FUL2-bound region at least partially overlapped the RIN-bound region. At the same time, 78% of the FUL1-bound region overlapped 73% of the FUL2-bound region. Additionally, 19% of the FUL1-bound region, 25% of the FUL2-bound region, and 3.2% of the RIN-bound region did not overlap the other transcription factor-bound regions. Analysis of the four typical binding sites, including the PAL type [CC(A/T)6GG], the N10 type [CTA(A/T)4TAG], and the typical RIN-binding motifs [C(C/T)(A/T)6(A/G)G and C(A/T)8G], showed that FUL1 and FUL2 exhibited DNA- binding specificities (Fujisawa et al. 2014).

Based on ChIP-chip and RNA-Seq analyses, it is known that FUL1 and FUL2 directly regulate many ripening-related genes that are also direct targets of RIN (Figure 3.11). These target genes, including *ACS2* and *ACS4*, ACC oxidase

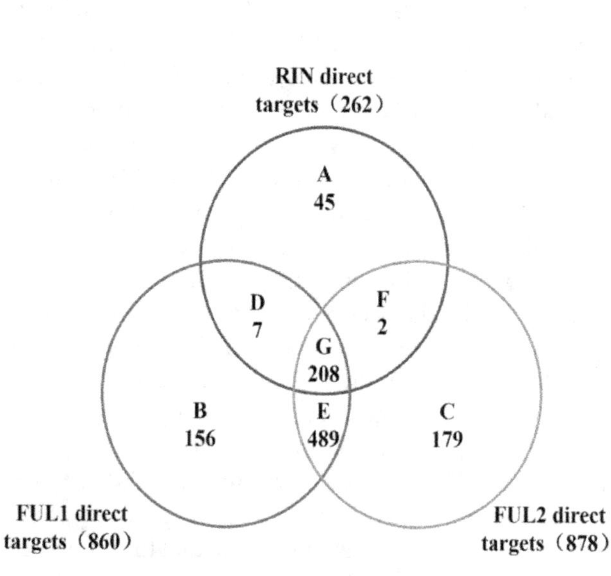

Figure 3.11 Venn diagram of target genes directly regulated by RIN, FUL1, and FUL2 in fruit ripening.

homolog *E8*, and other ripening-related key genes such as *AP2a*, *NOR*, and *RIN*, are involved in the ethylene pathway of fruit ripening. FUL1, FUL2, and RIN exhibit overlapping binding regions for many ripening-related genes, indicating that they regulate the expression of these target genes by forming protein complexes. However, some ripening-related genes have been found to be under the control of only one transcription factor. For example, a HD-zip homolog could only be bound by FUL1. Genes bound only by FUL2 have not been found. PSY1 can be bound by FUL1 and RIN but not by FUL2. In addition, PG2a, E4, and CNR are not directly bound by FUL1 and FUL2, but their transcription levels are highly correlated.

3.5.2.3 The Effect of Tomato TAGL1 on Fruit Ripening

TAGL1 is also a MADS-box family transcription factor; it can interact with RIN and FUL2 proteins but not with FUL1 in vitro (Leseberg et al. 2008; Fujisawa et al. 2014). Expression of the *TAGL1* gene increases during tomato fruit ripening; inhibition of *TAGL1* expression leads to abnormal pigment accumulation in the tomato fruit, reduced levels of carotenoids and ethylene, suppression of chlorophyll breakdown, and downregulation of many ripening-related genes. *TAGL1* overexpression promotes the accumulation of lycopene in the fruit and results in swollen sepals and increased levels of the yellow flavonoid naringenin chalcone. *TAGL1* is involved in the phenylpropanoid pathway and in yellow flavonoid biosynthesis. Additionally, TAGL1 can activate *ACS2* expression to induce ethylene biosynthesis and plays important regulatory roles in fruit ripening (Itkin et al. 2009; Vrebalov et al. 2009).

3.5.2.4 The Effect of Tomato MADS1 on Fruit Ripening

Tomato *MADS1* is highly expressed in sepals and fruits, and its expression increases during sepal development. However, *MADS1* expression decreases during fruit ripening. Silencing of *MADS1* expression using RNAi resulted in shorter ripening time, increased accumulation of carotenoids, and upregulation of genes including *PSY1*. The expression of genes related to ethylene biosynthesis, including *ACO1*, *ACS6*, *ACS1A*, and *ACO3* and the ethylene-responsive genes *E4* and *E8*, was upregulated, and ethylene production was increased two- to fourfold. *MADS1*-silenced etiolated tomato seedlings also had shorter hypocotyls and were more sensitive to ACC (Dong et al. 2013).

3.5.2.5 The Role of Other Tomato MADS-box Transcription Factors in Formation of the Pedicel Abscission Zone (AZ) and Fruit Ripening

In the middle portion of the tomato pedicel, there is a distinct region called the abscission zone (AZ). In unfavorable external environments, after failure of pollination, or during fruit ripening, the middle lamella between cells in the AZ is degraded, leading to the shedding of flowers and leaves. The formation of the pedicel AZ is completely inhibited in tomato *j* mutants; in these mutants, AZ cannot be formed. It was found that the *j* mutant was caused by the loss of function of *JOINTLESS*, a MADS-box gene, indicating that *JOINTLESS* is a key gene in the formation of the AZ in tomato (Mao et al. 2000). In tomato *rin* mutants, which have unripened fruits and enlarged sepals, the formation of the pedicel AZ is also inhibited to a certain degree. In 2012, Nakano et al. found that suppression of the MADS-box gene *MACROCALYX* (*MC*) using an *MC* antisense construct resulted in lack of AZ formation in the pedicels (Nakano et al. 2012). Moreover, MC protein could interact with JOINTLESS protein, and together, they regulated the expression of genes involved in AZ development such as *WUSCHEL* (*WUS*), *GOBLET* (*GOB*), and *LATERAL SUPPRESSOR* (*Ls*). These results clarified the molecular mechanism through which pedicel AZ formation is inhibited in *rin* mutants.

Overexpression and silencing of *FYFL* in transgenic tomatoes showed that the *FYFL* gene may regulate tomato fruit ripening by regulating the expression of ethylene-related genes. Transgenic tomatoes with *FYFL* overexpression (*35S:FYFL*) exhibited delayed ripening time, reduced accumulation of carotenoids, and increased storability and fruit hardness as well as reduced ethylene production. On the other hand, transgenic tomatoes in which *FYFL* had been silenced showed opposite phenotypes with respect to the fruit parameters mentioned above. Fruit-ripening-related genes, carotenoid biosynthetic genes (*PSY1*, *PDS*, and *ZDS*), genes involved in cell wall metabolism (*PG*, *Exp 1*, *XTH5*, and *TBG4*), and genes involved in fatty acid metabolism (*LoxA*, *LoxB*, and *LoxC*) were significantly downregulated in *35S:FYFL* fruit and significantly upregulated in *FYFL*-silenced fruit. Ethylene biosynthetic genes (*RIN*, *ACS2*, *ACO1*, and *ACO3*) and ethylene-responsive genes (*E4* and *E8*) were significantly upregulated in FYFL-silenced fruit and significantly downregulated in *35S:FYFL* fruit. Fruit hardness measurements showed that *35S:FYFL* fruit had increased hardness compared to wild-type fruits at the same developmental stages, while fruit hardness was significantly lower in *FYFL*-silenced fruits than in wild-type fruits. Storability testing showed that *35S:FYFL* fruit had significantly increased storability compared to wild type, while *FYFL*-silenced tomato had significantly decreased storability compared to wild type. In addition, ethylene production in *35S:FYFL* fruit was only 40%–50% of that in wild-type fruit, while ethylene production in *FYFL*-silenced fruit was 3.5-fold higher. This result indicates that *FYFL* may act upstream of ethylene to regulate tomato fruit ripening. *FYFL* is also involved in ethylene-related AZ development. Transgenic tomatoes overexpressing the *SlFYFL* gene exhibited a phenotype of delayed fruit abscission, while those with silenced SlFYFL had early fruit abscission. Microscopic observation showed that the formation of AZ cells was delayed in *35S:FYFL* tomato and that it occurred significantly earlier in *FYFL*-silenced tomato pedicel AZ. Quantitative measurement of the force needed to break the fruit stalk at the AZ showed that the break strength of *35S:FYFL* pedicels was significantly greater than that of wild-type pedicels, while that of *FYFL*-silenced pedicels was lower. Genes related to AZ development (*MC*, *JOINTLESS*, *Bl*, *WUS*, *Ls*, and *GOB*) had significantly downregulated expression in *35S:FYFL* pedicel AZ and significantly upregulated expression in *FYFL*-silenced pedicel AZ. Ethylene biosynthetic genes (*ACS4*, *ACS2*, *ACS1A*, *ACO1*, and *ACO3*) had significantly downregulated expression in *35S:FYFL* pedicel AZ and significantly upregulated expression in *FYFL*-silenced pedicel AZ. This finding indicates that the *FYFL* gene may regulate pedicel AZ development by regulating the expression of genes related to AZ development via ethylene (Xie et al. 2014).

It is likely that the *FYFL* gene also regulates the chlorophyll content of leaves via regulating the expression of ethylene-related genes to further regulate tomato leaf senescence. *35S:FYFL* leaves exhibited delayed senescence, increased chlorophyll content, significantly upregulated expression of genes related to photosynthesis (*Slcab7* and *SlrbcS*), and significantly downregulated expression of ethylene biosynthetic genes (*ACO1*, *ACO3*, *ACS1A*, *ACS2*, and *ACS6*). *FYFL*-silenced tomato leaves showed phenotypes opposite to those of *35S:FYFL* leaves (Xie et al. 2014).

MADS-box proteins usually function by forming protein complexes. Studies have shown that FYFL can interact with itself, with tomato fruit-ripening-related RIN, and with MADS1 protein. This finding indicates that FYFL may affect ethylene production by interacting with RIN and MADS1 to further regulate tomato fruit ripening. FYFL can also interact with JOINTLESS, which is related to the development of the AZ in tomato. The fact that JOINTLESS can interact with MC, which is also related to AZ development, indicates that FYFL can regulate AZ formation by forming protein complexes with JOINTLESS and MC to regulate fruit abscission in tomato (Xie et al. 2014).

3.5.2.6 Studies of the Regulation of MADS-box Transcription Factors in Ripening in Banana

Broadly speaking, banana includes banana and plantain, which are the most important food crops in tropical and subtropical developing countries. The economic value of bananas is high. First, they are a daily food source for approximately 400 million people worldwide and are listed as the fourth largest food crop by the Food and Agriculture Organization of the United Nations. Second, banana trading is the primary method through which banana-producing countries and regions obtain foreign currency. Third, bananas possess extremely high nutritional value; they are rich in vitamins A, C, and B6 and contain large amounts of calcium, potassium, and phosphorus. To ensure that banana fruits can be safely transported from production sites to the market, bananas are usually harvested green in commercial production. From the perspective of edibility, green bananas have not yet ripened; they need to go through postharvest ripening to reach the best quality for consumption. Banana is a typical climacteric fruit; its postharvest ripening process involves many physiological and biochemical changes, including the conversion of starch to sugar, the degradation of structural carbohydrates and polyphenols, an increase in the levels of isoamyl acetate, and chlorophyll degradation. These changes directly affect fruit qualities such as hardness, astringency, aroma, color, and postharvest shelf life, qualities that further affect the commercial values of bananas. Therefore, banana postharvest ripening is the core of studying banana fruit development. Artificially controlling banana postharvest ripening plays an important role in improving banana quality and extending shelf life; it also provides important assurance for countries developing banana production and increases their economic benefits.

MADS-box genes in banana were first reported in 2007. Inaba et al. isolated three MADS-box genes: MA-*MADSl*, MA-*MADS2*, and MA-*MADS3* (GenBank accession Nos. AY941798, AY941799, and AY941800, respectively). This study showed that of the three banana MADS genes, MA-*MADS2* had the highest expression in the peel and the lowest in the pulp. 1-MCP significantly inhibited MA-*MADS2* expression in peel tissue but had very little inhibitory effect on MA-*MADS2* expression in pulp tissue. MA-*MADS1* and MA-*MADS3* were expressed constitutively in peel and pulp tissues, independent of 1-MCP treatment (Inaba et al. 2007). A subsequent study found that Ma*MADS1*, Ma*MADS2*, Ma*MADS3*, and Ma*MADS4* possessed functions similar to those of the *SEP* genes in tomato fruit (Elitzur et al. 2010). Ma*MADS5* is an *AG* gene with functions similar to those of tomato *TAGL1* (Itkin et al. 2009; Vrebalov et al. 2009). Ma*MADS5* binds to the CArG box, and it can interact with a protein from the OVATE family that inhibits cell elongation. Inhibition of expression of the Ma*MADS1*, Ma*MADS2*, Ma*MADS3*, Ma*MADS4*, and Ma*MADS5* genes can affect the ripening process and cause tissue-specific effects in pulp and peel tissues. Of these genes, Ma*MADS1* is the most closely related to ripening. With the exception of Ma*MADS2*, the expression of these genes in pulp and peel tissue is induced by ethylene. Ma*MADS1* and Ma*MADS2* are homologs of the tomato *RIN* gene. Suppressing the expression of either Ma*MADS1* or Ma*MADS2* by RNAi resulted in specific ripening delay and extended shelf life as indicated by delayed color development and softening. It has been shown that silencing of Ma*MADS1* or Ma*MADS2* inhibits the climacteric respiration of banana and reduces ethylene biosynthesis. In the lines that showed the most suppression, ethylene production was almost undetectable, and the ripening process was dramatically delayed. However, transgenic banana fruits could respond to external ethylene, unlike tomato *rin* mutants. This result suggests that although ripening-related MADS-box genes are functionally conserved during fruit ripening, the regulatory functions of MADS-box genes in banana differ in some ways from those of tomato (Elitzur et al. 2016). A recent study found that *MaMADS36* interacts with 74 other proteins to form a co-expression network and could act as an important switch to regulate ethylene biosynthesis, starch degradation, softening, and sugar accumulation during banana ripening (Liu et al. 2021).

3.5.2.7 Studies of the Regulation of MADS-box Transcription Factors in Ripening in Other Fruit

In recent years, the key regulatory role of MADS-box transcription factors on fruit ripening has been validated in some other fruits. In apple fruit, suppression of the SEP1/2-like genes *MdMADS8* and *MdMADS9* resulted in sepaloid petals and greatly reduced fruit flesh development and ripening (Ireland et al. 2013). Silencing of *PrpSEP1* in peach delayed fruit ripening and softening (Li et al. 2017). In *SEP1*-silenced peach fruit, the expression levels of the ripening- and softening-related genes, *ACS2, EIN2, PME1, Endo-PG3, ACO1, ETR2, β-Hex2* and *Lox1*, were significantly reduced, while the transcription of *EXP 2* and *PME3* were significantly enhanced except at the harvest stage (Li et al. 2017). In strawberry fruit, *FaMADS1a* act as a negative regulator involved in fruit-ripening process, where overexpression of *FaMADS1a* delayed fruit ripening and inhibited the expressions of anthocyanidin-related genes *FaPAL6, FaCHS, FaDFR*, and *FaANS* (Lu et al. 2018). Qi et al. (2020) demonstrated that silencing *PaMADS7* using TRV-mediated virus-induced gene silencing inhibited fruit ripening and PaMADS7 could be involve in sweet cherry fruit softening by directly binding to the promoter of *PaPG1*.

In recent years, relying on the model plant tomato, the revealing mechanistic complexity and unique regulatory features of MADS-box family TFs on fruit ripening have been revealed. However, it is still limited, and the members of the MADS-box family involved in fruit ripening and their regulatory modes need to be further supplemented. In addition, current functional characterization of MADS-box family TFs has been limited to a few fruit species, mainly tomato, which limits knowledge of the role of MADS-box in other non-model plants. Given that the ripening process in non-climacteric fruits is ethylene-independent, whether there are differences in the regulation of MADS-box family TFs on the ripening process between climacteric and non-climacteric fruit needs to be further explored. More extensive bioinformatic interpretation and explicit functional validation will help us better understand the role of MADS-box family TFs in fruit ripening, which can be used as targets for genetic engineering to improve fruit quality in the future.

References

Alvarez-Buylla, E.R., Pelaz, S., Liljegren, S.J. et al. (2000). An ancestral MADS-box gene duplication occurred before the divergence of plants and animals. *PNAS* 97: 5328–5333.

Angenent, G., Franken, J., Busscher, M. et al. (1995). A novel class of MADS box genes is involved in ovule development in petunia. *Plant Cell* 7: 1569–1582.

Arora, R., Agarwal, P., Ray, S. et al. (2007). MADS-box gene family in rice: genome-wide identification, organization and expression profiling during reproductive development and stress. *BMC Genom* 8: 242.

Becker, A. and Theissen, G. (2003). The major clades of MADS-box genes and their role in the development and evolution of flowering plants. *Mol Phylogenet Evol* 29: 464–489.

Bemer, M., Karlova, R., Ballester, A.R. et al. (2012). The tomato FRUITFULL homologs TDR4/FUL1 and MBP7/FUL2 regulate ethylene-independent aspects of fruit ripening. *Plant Cell* 24: 4437–4451.

Benfey, P., Ren, L., and Chua, M. (1990). Tissue-specific expression from CaMV35S enhancer subdomains in early stages of plant development. *EMBO J* 9: 1677–1684.

Castillejo, C., Romera-Branchat, M., and Pelaz, S. (2005). A new role of the Arabidopsis SEPALLATA3 gene revealed by its constitutive expression. *Plant J* 43: 586–596.

Causier, B., Kieffer, M., and Davies, B. (2002). Plant biology. MADS-box genes reach maturity. *Science* 296 (5566): 275–276.

Coen, E. and Meyerowitz, E. (1991). The war of the whorls: genetic interactions controlling flower development. *Nature* 353: 31–37.

Colombo, L., Franken, J., Koetje, E. et al. (1995). The petunia MADS box gene FBPI7 determines ovule identity. *Plant Cell* 7: 1859–1868.

De, B.S., Raes, J., Van, D.P.Y., and Theissen, G. (2003). And then there were many: MADS goes genomic. *Trends Plant Sci* 8: 475–483.

de Folter, S. and Angenent, G. (2006). trans meets cis in MADS science. *Trends Plant Sci* 11: 224–231.

Díaz-Riquelme, J., Lijavetzky, D., Martínez-Zapater, J., and Carmona, M.J. (2009). Genome-wide analysis of MIKCC-type MADS box genes in grapevine. *Plant Physiol* 149: 354–369.

Ditta, G., Pinyopich, A., Robles, P et al. (2004). The SEP4 gene of Arabidopsis thaliana functions in floral organ and meristem identity. *Curr Biol* 14 (21): 1935–1940.

Dong, T., Hu, Z., Deng, L. et al. (2013). A tomato MADS-box transcription factor, SlMADS1, acts as a negative regulator of fruit ripening. *Plant Physiol* 163: 1026–1036.

Elitzur, T., Vrebalov, J., Giovannoni, J. et al. (2010). The regulation of MADS-box gene expression during ripening of banana and their regulatory interaction with ethylene. *J Exp Bot* 61: 1523–1535.

Elitzur, T., Yakir, E., Quansah, L. et al. (2016). Banana MaMADS transcription factors are necessary for fruit ripening and molecular tools to promote shelf-life and food security. *Plant Physiol* 171: 380–391.

Eriksson, E.M., Bovy, A., Manning, K. et al. (2004). Effect of the colorless non-ripening mutation on cell wall biochemistry and gene expression during tomato fruit development and ripening. *Plant Physiol* 136 (4): 4184–4197.

Favaro, R., Pinyopich, A., Battaglia, R. et al. (2003). MADS-box protein complexes control carpel and ovule development in Arabidopsis. *Plant Cell* 15: 2603–2611.

Fraser, P.D., Enfissi, E.M., Halket, J.M. et al. (2007). Manipulation of phytoene levels in tomato fruit: effects on isoprenoids, plastids, and intermediary metabolism. *Plant Cell* 19: 3194–3211.

Fujisawa, M., Nakano, T., Shima, Y., and Ito, Y. (2013). A large-scale identification of direct targets of the tomato MADS box transcription factor RIPENING INHIBITOR reveals the regulation of fruit ripening. *Plant Cell* 25: 371–386.

Fujisawa, M., Shima, Y., Higuchi, N. et al. (2012). Direct targets of the tomato-ripening regulator RIN identified by transcriptome and chromatin immunoprecipitation analyses. *Planta* 235: 1107–1122.

Fujisawa, M., Shima, Y., Nakagawa, H. et al. (2014). Transcriptional regulation of fruit ripening by tomato FRUITFULL homologs and associated MADS box proteins. *Plant Cell* 26: 89–101.

Giovannoni, J.J. (2004). Genetic regulation of fruit development and ripening. *Plant Cell* 16: S170–S180.

Giovannoni, J.J. (2007). Fruit ripening mutants yield insights into ripening control. *Curr Opin Plant Biol* 10: 283–289.

Immink, R., Gadella, T., Ferrario, S. et al. (2002). Analysis of MADS box protein-protein interactions in living plant cells. *PNAS* 99: 2416–2421.

Inaba, A., Liu, X., Yokotani, N. et al. (2007). Differential feedback regulation of ethylene biosynthesis in pulp and peel tissues of banana fruit. *J Exp Bot* 58: 1047–1057.

Ireland, H.S., Yao, J.L., Tomes, S. et al. (2013). Apple SEPALLATA1/2-like genes control fruit flesh development and ripening. *Plant J* 73 (6): 1044–1056.

Itkin, M., Seybold, H., Breitel, D. et al. (2009). TOMATO AGAMOUS-LIKE1 is a component of the fruit ripening regulatory network. *Plant J* 60: 1081–1095.

Ito, T. (2019). Functional dissection of class C genes in rice. *Plant Cell Physiol* 60 (6): 1164–1165.

Ito, Y., Kitagawa, M., Ihashi, N. et al. (2008). DNA-binding specificity, transcriptional activation potential, and the rin mutation effect for the tomato fruit-ripening regulator RIN. *Plant J* 55: 212–223.

Ito, Y., Nishizawa-Yokoi, A., Endo, M. et al. (2017). Re-evaluation of the rin mutation and the role of RIN in the induction of tomato ripening. *Nat Plants* 3: 866–874.

Ito, Y., Nishizawa-Yokoi, A., Endo, M. et al. (2015). CRISPR/Cas9-mediated mutagenesis of the RIN locus that regulates tomato fruit ripening. *Biochem Biophys Res Commun* 467: 76–82.

Jaakola, L., Poole, M., Jones, M. et al. (2010). A SQUAMOSA MADS box gene involved in the regulation of anthocyanin accumulation in bilberry fruits. *Plant Physiol* 153: 1619–1629.

Kater, M., Dreni, L., and Colombo, L. (2006). Functional conservation of MADS-box factors controlling floral organ identity in rice and Arabidopsis. *J Exp Bot* 57: 3433–3444.

Kaufmann, K., Melzer, R., and Theissen, G. (2005). MIKC-type MADS-domain proteins: structural modularity, protein interactions and network evolution in land plants. *Gene* 347: 183–198.

Kieffer, M. and Davies, B. (2001). Developmental programmes in floral organ formation. *Semin Cell Dev Biol* 12: 373–380.

Kitazawa, Y., Iwabuchi, N., Maejima, K. et al. (2022). A phytoplasma effector acts as a ubiquitin-like mediator between floral MADS-box proteins and proteasome shuttle proteins. *Plant Cell* 34 (5): 1709–1723.

Lai, X., Daher, H., Galien, A. et al. (2019). Structural basis for plant MADS transcription factor oligomerization. *Comput Struct Biotechnol J* 17: 946–953.

Leseberg, C.H., Eissler, C.L., Wang, X. et al. (2008). Interaction study of MADS-domain proteins in tomato. *J Exp Bot* 59: 2253–2265.

Li, J., Li, F., Qian, M. et al. (2017). Characteristics and regulatory pathway of the PrupeSEP1 SEPALLATA gene during ripening and softening in peach fruits. *Plant Sci* 257: 63–73.

Liljegren, S.J., Ditta, G.S., Eshed, Y. et al. (2000). SHATTERPROOF MADS-box genes control seed dispersal in Arabidopsis. *Nature* 404 (6779): 766–770.

Li, S., Chen, K.S., and Grierson, D. (2019). A critical evaluation of the role of ethylene and MADS transcription factors in the network controlling fleshy fruit ripening. *New Phytol* 221: 1724–1741.

Li, S., Xu, H.J.L., Ju, Z. et al. (2018). The RIN-MC fusion of MADS-box transcription factors has transcriptional activity and modulates expression of many ripening genes. *Plant Physiol* 176: 891–909.

Li, S., Zhu, B.Z., Pirrello, J. et al. (2020). Roles of RIN and ethylene in tomato fruit ripening and ripening-associated traits. *New Phytol* 226: 460–475.

Lin, Z., Hong, Y., Yin, M. et al. (2008). A tomato HD-Zip homeobox protein, LeHB-1, plays an important role in floral organogenesis and ripening. *Plant J* 55: 301–310.

Liu, J., Liu, M., Wang, J. et al. (2021). Transcription factor MaMADS36 plays a central role in regulating banana fruit ripening. *J Exp Bot* 72 (20): 7078–7091.

Lopez-Dee, Z.P., Wittich, P., Enrico, P.M. et al. (1999). OsMADS13, a novel rice MADS-box gene expressed during ovule development. *Dev Genet* 25: 237–244.

Lu, W., Chen, J., Ren, X. et al. (2018). One novel strawberry MADS-box transcription factor FaMADS1a acts as a negative regulator in fruit ripening. *Sci Hortic* 227: 124–131.

Ma, H. (2005). Molecular genetic analyses of microsporogenesis and microgametogenesis in flowering plants. *Annu Rev Plant Biol* 56: 393–434.

Manning, K., Tör, M., Poole, M. et al. (2006). A naturally occurring epigenetic mutation in a gene encoding an SBP-box transcription factor inhibits tomato fruit ripening. *Nat Genet* 38: 948–952.

Mao, L., Begum, D., Chuang, H. et al. (2000). JOINTLESS is a MADS-box gene controlling tomato flower abscission zone development. *Nature* 406: 910–913.

Martel, C., Vrebalov, J., Tafelmeyer, P., and Giovannoni, J.J. (2011). The tomato MADS-box transcription factor RIPENING INHIBITOR interacts with promoters involved in numerous ripening processes in a COLORLESS NONRIPENING-dependent manner. *Plant Physiol* 157: 1568–1579.

Messenguy, F. and Dubois, E. (2003). Role of MADS box proteins and their cofactors in combinatorial control of gene expression and cell development. *Gene* 316: 1–21.

Michaels, S.D., Ditta, G., Gustafson-Brown, C. et al. (2003a). AGL24 acts as promoter of flowering in Arabidopsis and is positively regulated by vernalization. *Plant J* 33: 867–874.

Michaels, S.D., He, Y., Scortecci, K.C., and Amasino, R.M. (2003b). Attenuation of *FLOWERING LOCUSC* activity as a mechanism for the evolution of summer-annual flowering behavior in Arabidopsis. *PNAS* 100: 10102–10107.

Morel, P., Heijmans, K., Ament, K. et al. (2018). The floral C-lineage genes trigger nectary development in Petunia and Arabidopsis. *Plant Cell* 30 (9): 2020–2037.

Mueller, C.G. and Nordheim, A. (1991). A protein domain conserved between yeast MCM1 and human SRF directs ternary complex formation. *EMBO J* 10: 4219–4229.

Nakano, T., Kimbara, J., Fujisawa, M. et al. (2012). MACROCALYX and JOINTLESS interact in the transcriptional regulation of tomato fruit abscission zone development. *Plant Physiol* 158: 439–450.

Nam, J., DePamphilis, C., Ma, H., and Nei, M. (2003). Antiquity and evolution of the MADS-box gene family controlling flower development in plants. *Mol Biol Evol* 20: 1435–1447.

Nam, J., Kim, J., Lee, S. et al. (2004). Type I MADS-box genes have experienced faster birth-and-death evolution than type II MADS-box genes in angiosperms. *PNAS* 101: 1910–1915.

Ocarez, N. and Mejía, N. (2016). Suppression of the D-class MADS-box AGL11 gene triggers seedlessness in fleshy fruits. *Plant Cell Rep* 35 (1): 239–254.

Otani, M., Aoyagi, K., and Nakano, M. (2020). Suppression of B function by chimeric repressor gene-silencing technology (CRES-T) reduces the petaloid tepal identity in transgenic Lilium sp. *PloS one* 15 (8): e0237176.

Parenicová, L., de Folter, S., Kieffer, M. et al. (2003). Molecular and phylogenetic analyses of the complete MADS-box transcription factor family in Arabidopsis: new openings to the MADS world. *Plant Cell* 15: 1538–1551.

Pelaz, S., Ditta, G., Baumann, E. et al. (2000). B and C floral organ identity functions require SEPALLATA MADS-box genes. *Nature* 405: 200–203.

Pinyopich, A., Ditta, G., Savidge, B. et al. (2003). Assessing the redundancy of MADS-box genes during carpel and ovule development. *Nature* 424: 85–88.

Pollock, R. (1990). A sensitive method for the determination of protein-DNA binding specificities. *Nucleic Acids Res* 18: 6197–6204.

Qi, X., Liu, C., Song, L., and Li, M. (2020). PaMADS7, a MADS-box transcription factor, regulates sweet cherry fruit ripening and softening. *Plant Sci* 301: 110634.

Qin, G., Wang, Y., Cao, B. et al. (2012). Unraveling the regulatory network of the MADS box transcription factor RIN in fruit ripening. *Plant J* 70: 243–255.

Royo, C., Torres-Pérez, R., Mauri, N. et al. (2018). The major origin of seedless grapes is associated with a missense mutation in the MADS-box gene VviAGL11. *Plant Physiol* 177 (3): 1234–1253.

Schwab, W., Davidovich-Rikanati, R., and Lewinsohn, E. (2008). Biosynthesis of plant-derived flavor compounds. *Plant J* 54: 712–732.

Seymour, G. (2013). Fruit development and ripening. *Annu Rev Plant Biol* 64: 219–241.

Shima, Y., Kitagawa, M., Fujisawa, M. et al. (2013). Tomato FRUITFULL homologues act in fruit ripening via forming MADS-box transcription factor complexes with RIN. *Plant Mol Biol* 82: 427–438.

Smaczniak, C., Muiño, J.M., Chen, D. et al. (2017). Differences in DNA binding specificity of floral homeotic protein complexes predict organ-specific target genes. *Plant Cell* 29 (8): 1822–1835.

Sommer, H., Beltran, J.P., Huijser, P. et al. (1990). Deficiens, a homeotic gene involved in the control of flower morphogenesis in Antirrhinum majus: the protein shows homology to transcription factors. *EMBO J* 9: 605–613.

Sugiyama, S.H., Yasui, Y., Ohmori, S. et al. (2019). Rice flower development revisited: regulation of carpel specification and flower meristem determinacy. *Plant Cell Physiol* 60 (6): 1284–1295.

Theiβen, G. and Saedler, H. (2001). Plant biology. Floral quartets. *Nature* 409: 469–471.

Tzeng, T.Y., Hsiao, C.C., Chi, P.J., and Yang, C.H. (2003). Two lily sepallata-like genes cause different effects on floral formation and floral transition in Arabidopsis. *Plant Physiol* 133: 1091–1101.

Vrebalov, J. (2002). A MADS-box gene necessary for fruit ripening at the tomato ripening-inhibitor (rin) locus. *Science* 296: 343–346.

Vrebalov, J., Pan, I., Arroyo, A. et al. (2009). Fleshy fruit expansion and ripening are regulated by the tomato SHATTERPROOF gene TAGL1. *Plant Cell* 21: 3041–3062.

Vrebalov, J., Ruezinsky, D., Padmanabhan, V., et al. (2002). A MADS-box gene necessary for fruit ripening at the tomato ripening-inhibitor (rin) locus. *Science* 296 (5566): 343–346.

Wang, R., Tavano, E.C.R., Lammers, M. et al. (2019). Re-evaluation of transcription factor function in tomato fruit development and ripening with CRISPR/Cas9-mutagenesis. *Sci Rep* 9 (1): 1–10.

Xie, Q., Hu, Z., Zhu, Z. et al. (2014). Overexpression of a novel MADS-box gene SlFYFL delays senescence, fruit ripening and abscission in tomato. *Sci Rep* 13: 4367.

Xu, H.Y., Li, G.X., Li, Q.Z. et al. (2004). Characterization of HoMADS1 and its induction by plant hormones during in vitro ovule development in Hyacinthus orientalis L. *Plant Mol Biol* 55: 209–220.

Yanofsky, M., Ma, H., Bowman, J. et al. (1990). The protein encoded by the Arabidopsis homeotic gene agamous resembles transcription factors. *Nature* 346: 35–39.

Yu, T., Tzeng, D.T.W., Li, R. et al. (2019). Genome-wide identification of long non-coding RNA targets of the tomato MADS box transcription factor RIN and function analysis. *Ann Bot* 123 (3): 469–482.

Zhang, S., Yao, J., Wang, L. et al. (2022). Role of grapevine SEPALLATA-related MADS-box gene VvMADS39 in flower and ovule development. *Plant J* 111: 1565–1579.

4

NAC Transcription Factor Family Regulation of Flesh Fruit Development and Ripening

4.1 Introduction

NAC (*NAM/ATAF1, 2/CUC2*) family transcription factors (TFs) are among the largest plant-specific TF families. NAC is the abbreviation of the first letters from *NO APICAL MERISTEM* (*NAM*); *ATAF1* and *ATAF2*; and *CUP-SHAPED COTYLEDON 2* (*CUC2*). The *NAM* gene was originally identified in *Petunia*, while *ATAF1*, *ATAF2*, and *CUC2* genes were originally found in *Arabidopsis*. Proteins from the NAC TF family contain the NAC domain, a highly conserved N-terminus, and a divergent C-terminus. The NAC TF family has been found to be present in the genomes of many plant species, including rice (*Oryza sativa*), *Arabidopsis thaliana*, tomato (*Solanum lycopersicum*), tobacco (*Nicotiana tabacum*), soybean (*Glycine* max L.), poplar (*Populus trichocarpa*), Rangpur lime (*Citrus limonia* Osbeck), and barley (*Hordeum vulgare* L.).

NAC family TFs play multiple roles in plant growth and development, environmental stress responses, and fruit ripening. Thus, they have become a research hotspot in plant gene functions and expression regulations.

This chapter first provides an overview of the origin, classification, structure, and functions of the plant NAC family TFs. Then, the biological functions of the NAC family TFs are introduced. Finally, the functions of NAC family TFs in fruit development and ripening are discussed.

4.2 Overview of the Plant NAC Family TFs

TFs are an important subject of postgenomic studies. As trans-acting factors, TFs participate in various signal transduction pathways; they specifically bind to the cis-acting elements and activate the expression of downstream target genes. A great number of TFs exist in plants. Among the 27,000 genes in *Arabidopsis thaliana*, 5.9% encode TFs (Riechmann et al. 2000).

Transcriptional Regulation of Flesh Fruit Development and Ripening, First Edition. Yunbo Luo.

NAC (NAM/ATAF1, 2/CUC2) family TFs are plant specific and widely distributed in land plants; they are among the largest TF families in plant genomes. The name of the NAC family originated from the *NO APICAL MERISTEM* (*NAM*) gene in *Petunia hybrida* and the *ATAF1*, *ATAF2*, and *CUP-SHAPED COTYLEDON 2* (*CUC2*) genes in *Arabidopsis thaliana* (Souer et al. 1996; Aida et al. 1997).

NAC family TFs contain a highly conserved N-terminus and a divergent C-terminus. The N-terminus contains a DNA-binding domain of approximately 150 amino acids and can be divided into five subdomains (A–E). Among them, the nuclear localization signal (NLS) is found in the C and D subdomains. Both the N-terminus and the C-terminus exhibit protein-binding activity (Zhu et al. 2014a, 2014b). NAC proteins can function as homo- or heterodimers; the dimerization is necessary for stable DNA binding (Olsen et al. 2005a).

Studies have shown that NAC proteins are involved in plant growth and development and play roles in signal transduction as well as abiotic and biotic stress responses. In recent years, accumulating evidence has confirmed the regulatory role of NAC TFs in fruit ripening.

4.2.1 Origin of the NAC Family TFs

NAC family TFs are plant-specific TFs discovered only recently and possess various biological functions. They are widely distributed in land plants. The very first NAC TF gene discovered is the *NO APICAL MERISTEM* (*NAM*) gene from *Petunia* identified through transposon tagging by Souer et al. in 1996. The *nam* mutants failed to develop the apical meristem; most seedlings died, and the few surviving seedlings exhibited abnormal development of floral organs. The *NAM* gene also plays a role in determining the positions of meristems and primordia (Souer et al. 1996).

In 1997, Aida et al. identified the *CUP-SHAPED COTYLEDON 2* (*CUC2*) gene in *Arabidopsis*, which produced similar effects as the *NAM* gene in *Petunia*. The *cuc2* mutants exhibited abnormal apical meristem development. Cotyledons of the mutant seedlings were fused into a cup shape; the sepals and stamens were also fused in the mutant plants (Aida et al. 1997). Because the *CUC2* gene, the *NAM* gene reported by Souer et al., and the *ATAF1* and *ATAF2* genes cloned by Hirt all encode proteins that contain a conserved amino acid sequence at the N-terminus, with approximately 150 amino acid residues, the first letters of their names were used to name this conserved domain the NAC domain. Proteins that contain the NAC domain are referred to as NAC proteins.

In 2004, Ernst et al. (2004) analyzed the NAC domain structure of the *Arabidopsis* ANAC019 using X-ray crystallography. The results showed that this NAC domain formed symmetrical homodimers; each monomer contained six antiparallel β-sheets and three α-helices. A short antiparallel β-sheet and hydrogen bonds/salt bridges connected the arginine and glutamate side chains in the monomers to form dimers (Yamasaki et al. 2008). Surprisingly, the four longer β-sheets in NAC protein monomers highly resemble the spatial structure of the four β-sheets in the WRKY domain in plant WRKY TFs . The WRKYGQK residues in a β-strand of the WRKY proteins form a concave face that is positively charged and responsible for DNA binding. The NAC domain contains a conserved WKATGXD[K/R] sequence that is similar to the WRKYGQR in the WRKY domain. The surface electrical charge distribution also indicates the potential for DNA binding. Thus, these two plant-specific TFs, NAC and WRKY, are evolutionarily close (Yamasaki et al. 2008). Furthermore, the glial cells missing (GCM) TFs in animals are highly homologous to WRKY TFs (Cohen et al. 2003). Among the WRKY, GCM, and NAC TFs, only members of the group I WRKY proteins are found in primitive eukaryotes, while GCM and NAC are only found in animals and plants, respectively. Therefore, Yamasaki et al. hypothesized the following evolutionary relationship among these three TFs: animal GCM and plant NAC WRKY TFs all originated from primitive eukaryotic WRKY proteins(Yamasaki et al. 2008), as shown in Figure 4.1.

Members of NAC family TFs containing the NAC domain have been discovered in many plant species, including *Arabidopsis*, rice (*Oryza sativa*), wheat (*Triticum aestivum*), barley (*Hordeum vulgare*), corn (*Zea mays*), potato (*Solanum tuberosum*), rapeseed (*Brassica campestris*), pumpkin (*Cucurbita moschata*), soybean (*Glycine* max), sugarcane (*Saccharum officinarum*), snapdragon (*Antirrhinum majus*), broad bean (*Vicia faba*), sweet orange (*Citrus sinensis*), peanut (*Arachis hypogaea*), tomato (*Solanum lycopersicum*), bell pepper (*Capsicum annuum*), mandarin orange (*Citrus reticulata*), white mangrove (*Avicennia marina*), and chickpea (*Cicer arietinum*) (Li et al. 2011b). There are 19,997 NAC family TFs from 150 species in the Plant Transcription Factor Database (http://planttfdb.cbi.pku.edu.cn), including 138 NAC family TF genes from *Arabidopsis*, 170 from rice, 269 from soybean, 280 from tobacco (*Nicotiana tabacum*), and 101 from tomato (Jin et al. 2014).

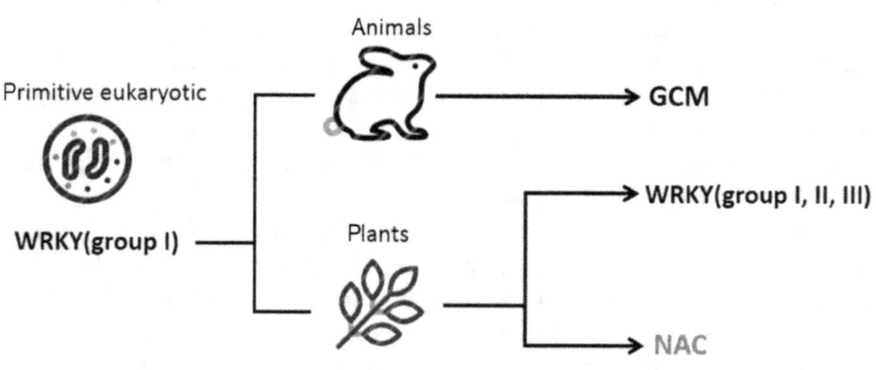

Figure 4.1 Phylogenetic tree of WRKY, GCM, and NAC family TFs.

4.2.2 Classification of NAC TFs

In 2003, Ooka et al. (Ooka et al. 2003) conducted a whole-genome analysis and found 105 NAC family TFs in *Arabidopsis* and 75 in rice. Based on the similarity of the NAC domain amino acid sequence, the NAC TF family members were divided into two large groups, Groups I and II, with a total of 18 subgroups. Group I was divided into 14 subgroups: TERN, ONAC022, SENU5, NAP, AtNAC3, ATAF2, OsNAC3, NAC2, ANAC011, TIP, OsNAC8, OsNAC7, NAC1, and NAM. Group II was divided into four subgroups: ANAC001, OsNAC003, ONAC001, and ANAC063. Subsequent phylogenic analyses of NAC family TFs in rice, poplar, soybean, and tobacco used the classification devised by Ooka as reference. Among the 18 subgroups, four subgroups consisted solely of *Arabidopsis* NAC TFs; two subgroups consisted of NAC TFs from monocot rice and wheat; and one subgroup was specific to Solanaceae (Hu et al. 2010; Nuruzzaman et al. 2010; de Oliveira et al. 2011).

A genome-wide analysis by Nuruzzam et al. compared the 151 rice NAC TF genes with 117 *Arabidopsis* NAC TF genes. The results suggested that different groups could possess different biological functions. For example, many stress-responsive genes were classified into *SNAC* (Nuruzzaman et al. 2010). Nakashima et al. conducted a further phylogenetic analysis and classified these abiotic stress-responsive *NAC* genes into six major functional groups: *NAM/CUC3*, *SND*, *TIP*, *SNAC*, *ANAC034*, and *ONAC4* (Nakashima et al. 2012).

In 2008, Rushton et al. performed a broader phylogenetic analysis of the *NAC* genes from *Arabidopsis*, rice, soybean, tomato, and tobacco. They proposed that there were seven subgroups in the NAC family TFs, where six subgroups were shared by plants from different families, while one subgroup was specific to Solanaceae. Therefore, this subgroup was named tobacco *NAC* genes (*TNACS*). The N-terminus of the TNACS proteins exhibits clear differences from that of other subgroups. However, there has been no study showing whether this structural difference is related to different functions (Rushton et al. 2008). As the sequence information of NAC family TFs continues to be updated, cluster analyses in a broader range will likely lead to new discoveries.

4.2.3 Localization of the NAC TFs

Most NAC TFs are localized in the nucleus, but a group of special NAC TFs, called the NAC membrane-bound TFs (NTLs), are membrane tethered and inactive under normal conditions. In the presence of hormone or environmental stimuli, NTLs can be released from the membrane and enter the nucleus to regulate gene expression (Kim et al. 2006; Wang et al. 2016). It has been shown that membrane-mediated transcriptional regulation is widely present (Seo et al. 2008). Whole-genome analysis predicted that there were 13 membrane-associated NAC TFs (Kim et al. 2007b), including 5 in rice (Olsen et al. 2005a) and 13 in tomato, named SlNACMTF1–SlNACMTF13 (Bhattacharjee et al. 2017).

NTM1 is the first membrane-bound NAC TF found in *Arabidopsis* in 2006. NTM1 can sense cytokinin and be released from the cell membrane via cysteine protease-mediated hydrolysis. Then, NTM1 enters the nucleus and negatively regulates cell division (Kim et al. 2006). At present, studies of membrane-bound NAC TFs are mostly concentrated in *Arabidopsis* (Kim et al. 2006, 2007b; Seo et al. 2010; Li et al. 2011a; De Clercq et al. 2013; Ng et al. 2013; Tang et al. 2016), while a few reports are available in rice (Guo et al. 2018), *Medicago* (Duan et al. 2017), maize (Wang et al. 2016), and soybean (Li et al. 2016b). These NTLs have been shown to participate in regulating plant stress responses (Kim et al. 2007a; Seo et al. 2010; De Clercq et al. 2013; Duan et al. 2017), cell division (Kim et al. 2006), mitochondrial retrograde signaling (Ng et al. 2013), auxin biosynthesis, and leaf margin development (Tang et al. 2016).

4.2.4 Structure of NAC TFs and the Mechanism of Their Functions

NAC TFs contain a highly conserved N-terminus, the NAC domain, which can bind DNA and other proteins. The NAC domain contains five subdomains (A, B, C, D, E); subdomains A, C, and D are highly conserved in different species, while the B and E subdomains are divergent but are both indispensable for the function of NAC TFs. The C and D subdomains contain NLSs and may be involved in nuclear localization and the recognition of specific cis-acting elements on the promoter, as shown in Figure 4.2. Thus, the A, C, and D subdomains are speculated to play roles in NAC protein functions (Ooka et al. 2003). Additionally, NAC TFs can bind other proteins via the NAC domain. For example, they can interact with the conserved RING family proteins, which regulate the transcriptional activity of NAC proteins in *Arabidopsis* (Greve et al. 2003). In 2004, Ernst et al. (Ernst et al. 2004) used X-ray crystallography to analyze the NAC domain in Arabidopsis ANAC019. The results revealed that the NAC domain forms a symmetrical homodimer; each monomer contains six antiparallel β-sheets and three α-helices, and a short antiparallel β-sheet and hydrogen bonds/salt bridges connect the arginine and glutamate side chains of the monomers into dimers (Yamasaki et al. 2008). This configuration is different from the classic helix-turn-helix structure; it is a novel fold structure for TFs—a few helices surrounding an antiparallel β-sheet. The NAC domain does not contain any known DNA-binding motif but

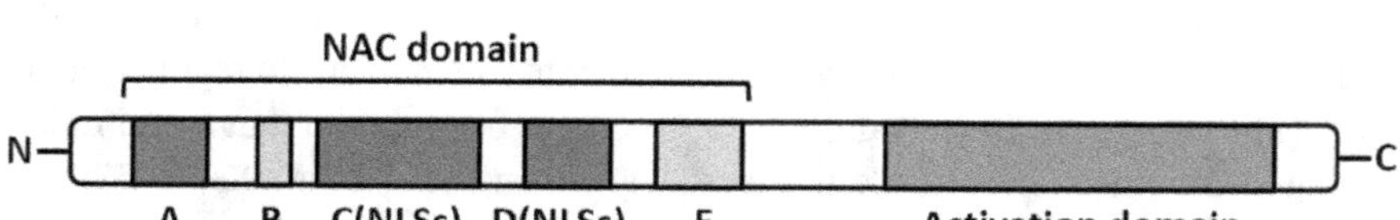

Figure 4.2 The structure of NAC proteins.

might form functional NAC protein homo- or heterodimers through salt bridges. The surface of this type of dimer is positively charged, which may be related to DNA binding. A study has shown that the poorly conserved C-terminus affects the stability of NAC protein dimers (Hegedus et al. 2003). The C-terminus of NAC TFs is highly divergent and is the transcriptional activation domain. Its common feature is that certain simple amino acids are present in high frequencies, such as serine, threonine, proline, and glutamate (Olsen et al. 2005a), which is the common characteristic of the transcriptional activation domain in plant TFs. Although the C-terminus is highly variable, there are 13 relatively conserved motifs. These motifs are distributed in NAC proteins in different subgroups; NAC proteins from the same subgroup contain the same motif, indicating that the NAC proteins from the subgroup might have similar biological functions (Ooka et al. 2003). Most recently, studies have found that the C-terminus of certain NAC proteins exhibit protein-binding activity and a transmembrane (TM) element. An analysis of the *Arabidopsis* genome found that at least 13 *Arabidopsis* NAC proteins have transmembrane functions and their C-terminus contains an α-helical TM sequence (TMS). Moreover, many NAC-MTFs exhibited upregulated gene expression under stress conditions, indicating that they may be involved in stress responses (Kim et al. 2008).

In *Petunia nam* mutants, the dTph1 transposon added six bases, which caused a 2-amino acid addition in the A subdomain. This change further affected the α-helix structure in the N-terminus and caused a loss of function of the NAM protein (Souer et al. 1996). In *Arabidopsis cuc1* mutants, the *cuc1-1* allele encoded a protein containing an amino acid substitution that altered the lysine-123 to a threonine in the B subdomain. This alteration may have affected the nuclear localization and DNA binding of the NAC protein, which caused the *cuc1* mutants to fail to form shoot apical meristem (Takada et al. 2001; Olsen et al. 2005b).

In the *Arabidopsis* genome, most *NAC* genes contain three exons and two introns. Usually, the first two exons encode the conserved NAC domain, while the third exon encodes the transcriptional activation domain. Additionally, a small fraction of *NAC* genes contain extra domains at the C-terminus and encode proteins with corresponding biochemical functions. In 2009, Kim et al. discovered at least 5 and 18 NAC TFs that are membrane-tethered transcription factors (MTTFs) in model plant species rice and *Arabidopsis*, respectively. The functions of four *Arabidopsis* membrane-bound NAC TFs (NTM1, NTL6, NTL8, and NTL9) have been studied. These TFs have similar structures, with an N-terminal cytoplasmic domain and C-terminal α-helical transmembrane domain (TMD). The cytoplasmic domain contains the transcription factor motif. NAC membrane-bound TFs are tethered on different membranes (NTM1 is tethered on the endoplasmic reticulum membrane, while NTL8 and NTL9 are located on the cytoplasm membrane). After receiving signals from external stimuli, the proteolytic cleavage releases the TF motif from the membrane, which allows the NAC TF to enter the nucleus and fulfill its biological functions (Li et al. 2011b).

Typical NAC TFs contain an N-terminal NAC domain and a divergent C-terminal transcription regulatory domain. In addition, there are atypical NAC TFs consisting of five main types as follows: the NAC domain contains a negative regulatory domain, and the C-terminal transcription regulatory domain contains a transmembrane motif; only the NAC domain is present; two tandem repeats of the NAC domain are present; an N-terminal extension region is present before the NAC domain; the NAC domain is at the C-terminus, while the transcription regulatory domain is at the N-terminus, with a conserved zinc finger motif in the middle with a DNA-binding function (Yang et al. 2014).

Most NAC proteins can form homo- or heterodimers, which may be the basic form for DNA binding (Olsen et al. 2004). Although many NAC proteins have been found to form homo- or heterodimers, there are exceptions. For example, the *Arabidopsis* AtNAC2 protein can form homodimers, but they cannot form heterodimers with AtNAC3 (He et al. 2005). In addition, NAC proteins can interact with other proteins. ANAC019, ANAC055, and ANAC072 can specifically bind to the CATGTG sequence in vitro or in vivo and activate the transcription of the *GUS* reporter gene. Meanwhile, ANAC019, ANAC055, and ANAC072 can all directly bind to the stress-inducible zinc finger homeodomain 1 (ZFHD1) protein, and this binding is necessary for the activation of downstream gene transcription (Tran 2004).

Because of the conserved basic amino acid residues at the N-terminus of the NAC protein, the surface is rich in positive charges, which may be directly involved in the binding of NAC proteins to DNA (Ernst et al. 2004). For example, a combination of K79A/R85A/R88A or R85A/R88A mutations could abolish the ability of the ANAC019 protein to bind DNA. In addition, a flexible loop is formed from K79-R85 between β1 and β2 of the ANAC019 protein; this sequence is highly conserved in the NAC TF family. Common DNA-binding elements are positively charged α-helices, but certain proteins can interact with DNA through the loop, while other proteins interact with DNA via β-sheets. Thus, it is possible that NAC proteins bind to DNA through both loops and β-sheets. Currently, the mechanism of the interactions between NAC proteins and DNA is not completely clear. However, many specific DNA sequences bound by NAC proteins have been isolated. *Arabidopsis* ANAC019, ANAC092, ANAC055, NAC072/RD26, NAC1, and AtNAM, as well as TaNAC in wheat (*Triticum aestivum*), may bind to the same DNA sequence—CGT[G/A]. *Arabidopsis* calmodulin-binding NAC protein (CBNAC) binds to the core GCTT sequence. These results indicate that not all NAC proteins recognize the same DNA sequence (Peng et al. 2010).

4.2.5 Regulation of the Expression of NAC TFs

The regulation of NAC TFs mainly involves the regulation of gene expression at the posttranscriptional level, protein modifications, and interactions with other proteins. The regulation of NAC TFs at the transcriptional level is primarily through miRNA.

4.2.5.1 The Regulation of NAC Gene Expression by miRNA

MicroRNAs (miRNAs) are noncoding single-stranded RNA molecules ~22 nt in length and encoded by endogenous genes. miRNAs are widely present in the genomes of animals and plants. Most miRNA genes exist as a single copy, multiple copies, or clusters. They are negative regulators of eukaryotic gene expression and function in three ways: mRNA cleavage at the posttranscriptional level, inhibition of translation, and binding inhibition of the target genes to regulate gene expression in plants, thereby regulating plant organ morphogenesis, growth and development, hormone secretion, signal transduction, and responses to environmental stresses. TFs are the main targets of miRNAs.

Studies have shown that miRNAs could regulate the expression level of *NAC* genes to control the biological functions of NAC TFs. In 2002, Rhoades et al. predicted that in *Arabidopsis*, miR164 was involved in regulating five *NAC* TF genes: *At5g61430*, *At5g07680*, *At1g56010/NAC1*, *At3g15170*, and *At5g53950/CUC2* (Rhoades et al. 2002). In 2007, Guo et al. found that miR164 could guide the cleavage of endogenous *NAC1* mRNA and produce 3′-specific fragments. When mutations occurred in the *NAC1* gene that disrupted the base pairing with miR164, the degradation of *NAC1* mRNA was blocked. In miR164 mutants, the expression level of miR164 was significantly reduced but the expression level of *NAC1* gene was significantly increased, which led to an increase in lateral root number. However, these mutant phenotypes could be restored by overexpressing *MIR164a* and *MIR164b*. In contrast, inducible expression of miR164 in wild-type (WT) plants caused a decreased *NAC1* expression and a reduced number of lateral roots (Guo et al. 2005). Laufs et al. found that overexpressing miR164 in *Arabidopsis* resulted in lowered expression levels of *CUC1* and *CUC2*, while *CUC3* was not affected. When the base pairing of miR164 and *CUC1* or *CUC2* was disrupted or miR164 was mutated, the boundary domains of *Arabidopsis* meristem were enlarged. These results indicate that miR164 constrains the expansion of the boundary domain by regulating the expression of *CUC1* and *CUC2* (Laufs et al. 2004). In 2005, Baker et al. found that in *Arabidopsis* eep1 mutants, which formed extra petals in early-arising flowers, there was a loss-of-function mutation of miR164c. Thus, miR164c controls petal number via regulating the expression of *CUC1* and *CUC2* (Baker et al. 2005).

4.2.5.2 Regulation of NAC TFs at the Protein Level

Phosphorylation is an important posttranslational modification of TFs. Posttranslational phosphorylation can not only regulate the cellular localization of TFs but also alter their activity. A study by Kaneda et al. showed that after induction by an avirulent pathogen recognition signal, the rice OsNAC4 protein was rapidly accumulated in the cytoplasm and phosphorylated. This protein translocated into the nucleus in a phosphorylation-dependent manner and regulated the expression of apoptosis-related genes, thereby regulating the hypersensitive response (HR) (Kaneda et al. 2009).

Protein degradation mediated by the ubiquitin/proteasome pathway is another important mechanism that regulates NAC TFs at the protein level. In *Arabidopsis*, the *NAC1* gene can transduce the auxin signal and promote lateral root development. SINAT5, a protein with the RING motif, has ubiquitin protein ligase activity and can ubiquitinate and degrade the NAC1 protein. Overexpressing *SINAT5* led to NAC protein degradation by ubiquitination and the loss of its biological functions, which abolished the ability of NAC1 to transduce auxin signal (Xie et al. 2002).

In addition, NAC TFs can interact with other proteins, which affects the activity of NAC TFs and their regulation of downstream gene expression. Tran et al. (2007) showed that *Arabidopsis* ANAC019, ANAC055, and ANAC072/RD26 can directly interact with zinc finger protein AFHD1 and activate the expression of the downstream gene *ERD1*.

4.3 NAC Family TFs Regulate Fruit Ripening

The NAC TF family is one of the largest TF families specific to plants. As research progresses, additional functions of the NAC TF family are discovered. The regulation of fruit ripening by the NAC family was discovered only recently, but an increasing number of studies have shown that NAC TF family members are involved in the regulation of ripening in fruits including tomato, banana, and papaya. How do NAC TFs regulate fruit ripening in plants? A detailed review of the regulation of fruit ripening by NAC TFs based on recent studies is provided below.

4.3.1 NAC Family TFs Regulate Tomato Fruit Ripening

The regulation of tomato fruit ripening by NAC TFs began with the study of *Non-ripening* (*nor*) mutants. The nor mutants were isolated by Kerr at the Horticultural Research Institute, Ontario, from the Italian Winter cultivar. The fruit-ripening process in the nor mutants was significantly inhibited; no climacteric respiration was present, with reduced lycopene production, increased fruit hardness, and slow fruit softening (Giovannoni 2007). Bioinformatic analysis showed that the

NOR gene belongs to the NAC TF family. The mutation in *NOR* significantly affected tomato fruit ripening, but there has been no report so far regarding the detailed molecular mechanism of how NOR regulates fruit ripening.

Zhu et al. further investigated the function of NAC family TFs other than *NOR* in the regulation of tomato fruit ripening. They found a new NAC family TF in tomato and named it *SlNAC4*. Analysis of the expression pattern of *SINAC4* in various tomato tissues showed that *SlNAC4* exhibited a higher expression level in sepals at the onset of fruit ripening. Various abiotic stresses (wounding, NaCl, drought, low temperature) can induce the expression of *SlNAC4*. In *SlNAC4*-RNAi transgenic tomato, fruit ripening was delayed by 2–3 days and chlorophyll breakdown was inhibited. Ethylene biosynthesis was reduced mainly by repressing the expression of ethylene biosynthetic genes of system II. Carotenoid accumulation was reduced by the alteration of the carotenoid pathway flux. Moreover, during later stages of fruit ripening, the placenta failed to mature. These results indicate that the *SlNAC4* gene positively regulates fruit ripening by affecting ethylene biosynthesis and the accumulation of carotenoids. Additionally, Zhu et al. found that the expression of *SlNAC4* was not induced by ethylene. It is possible that *SlNAC4* functions upstream of the ripening regulator *RIN* and positively regulates *RIN*. A yeast two-hybrid experiment further showed that the SlNAC4 protein can interact with RIN and NOR proteins. The above results indicate that the tomato *SlNAC4* gene can regulate ethylene-dependent and ethylene-independent fruit ripening, from which the function of *SINAC4* in the regulatory network of tomato fruit ripening can be predicted (Zhu et al. 2014b), as shown in Figure 4.3.

In 2013, Ma et al. cloned a NAC TF gene from tomato leaf and named it *SlNAC1*. It has been found that *SlNAC1* plays a regulatory role in tomato fruit ripening through regulating the biosynthesis of ethylene and ABA. Overexpressing *SlNAC1* could increase the cold tolerance of tomato plants. Transient expression of the SlNAC1::GFP fusion protein in onion epidermal cells was performed, and confocal microscopy observation revealed that the SlNAC1::GFP fusion protein was localized in the nucleus. Overexpression and antisense constructs were generated by inserting the coding region of *SlNAC1* into the pBI121 vector and transformed into tomato via the *Agrobacterium tumefaciens*-mediated leaf disc method. PCR, qRT-PCR, and western blotting were used to detect transgenic plants with kanamycin resistance. The results showed that transgenic plants overexpressing *SlNAC1* and antisense repressing *SlNAC1* expression were obtained. Compared to the WT, the overexpression line had mature fruits yellow or orange in color, a decreased number of carpels, reduced thickness of the pericarp, and enlarged and more abundant seeds. In contrast, antisense lines had deep red mature fruit, reduced numbers of carpels and seeds, a thickened pericarp, and smaller seeds. High-performance liquid chromatography (HPLC) analysis showed that the level

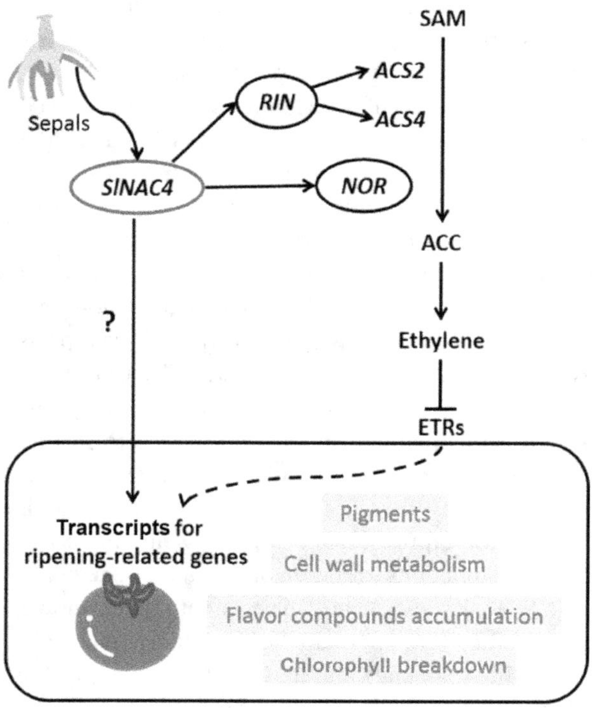

Figure 4.3 Model of suspected functions of *SlNAC4* in the regulatory network of fruit ripening.

of lycopene decreased in the fruit of the overexpression line, while the contents of lutein and β-carotene increased. The lycopene level in fruit of the antisense line increased. Gas chromatography analysis showed that the ethylene level in the fruit of the overexpressing line was significantly lower than that in the WT, whereas fruit softening was earlier. In the fruit of the antisense line, the release of ethylene was delayed, the climacteric peak was lower than that of the WT, and fruit softening was delayed. qRT-PCR results showed that compared to the WT, the overexpression line exhibited downregulated expression of lycopene biosynthesis-related genes in the fruit, while downstream of lycopene degradation, lutein and β-carotene synthesis-related genes were upregulated. Ethylene biosynthesis and signal transduction-related genes were downregulated, and genes related to cell wall metabolism were upregulated. In the fruits of the antisense line, the expression of the genes described above was opposite to that of the overexpression line. The microarray results were consistent with those from qRT-PCR. Yeast one-hybrid analysis showed that SlNAC1 directly regulates the expression of *ACS2*, *ACS4*, *ACO1*, and *PSY1*. Compared to the WT, the overexpression line exhibited increased ABA content in the fruit, while the antisense line exhibited decreased ABA content. Compared to untreated fruit at the same stage, the overexpression line at the MG stage treated with NDGA resulted in increased fruit firmness and delayed fruit ripening. The fruit of the antisense line at the MG stage treated with ABA was characterized by reduced firmness and earlier ripening. qRT-PCR analysis showed that the key genes in ABA biosynthesis, *NCED1* and *NCED2*, were upregulated in the overexpression line but downregulated in the antisense line (Ma 2013; Ma et al. 2014).

In 2016, to further study the function of *SlNAC1* in tomato fruit ripening, Meng et al. constructed an antisense construct and transformed it into tomato. They found that *SlNAC1*-RNAi transgenic tomato fruit exhibited delayed ripening, a decreased level of endogenous ABA, slower softening, and a deeper red appearance with increased carotenoids and lycopene contents when fruits were fully ripened. RT-PCR results showed that in *SlNAC1*-RNAi transgenic tomato fruits, genes related to lycopene and ethylene biosynthesis were upregulated, whereas genes related to cell wall metabolism and ABA biosynthesis were downregulated (Meng et al. 2016).

In 2016, to study the functions of NAC family TFs in tomato fruit ripening, Kou et al. predicted the secondary and tertiary structures of two NAC family TFs, SNAC4 (also known as SlNOR-like1, SlNAC3) and SNAC9 (SlNAC19, SlNAP2). They found that the tertiary structure of SNAC9 was similar to that of ATNAP (proven to play important roles in fruit senescence and ethylene production). The biofunction prediction results showed that SNAC4 and SNAC9 could be involved in the signal transduction of many plant hormones and senescence processes. Kou et al. applied VIGS to silence *SNAC4* and *SNAC9* and investigated the effects of these two genes on tomato fruit ripening. The results showed that the silencing of *SNAC4* and *SNAC9* significantly repressed the expression of *LeACS2*, *LeACS4*, and *LeACO1*, thereby inhibiting the process of fruit ripening. *SNAC4* silencing also led to the repression of genes related to ethylene signal transduction, except for *LeERF2*. Further study showed that the silencing of *LeACS4*, *LeACO1*, and *LeERF2* could also inhibit the expression of *SNAC4* and *SNAC9*, while silencing the NR gene could induce the expression of these genes. SNAC4 can also directly bind to the promoters of *SlCYP707A1* and *SAPK3*, activate *SAPK3* but inhibit SlCYP707A1. SNAC9 can interact with SlpYL9 and SlaREB1 (which are key proteins in the ABA signaling pathway) (Yang et al. 2021). The results of the above studies have shown that *SNAC4* and *SNAC9* may regulate tomato fruit ripening by simultaneously affecting ethylene and ABA metabolism and signal transduction (Kou et al. 2016; Yang et al. 2021).

In 2018, Gao et al. used VIGS to screen the NAC family members that had high expression levels in tomato fruit and found that silencing *NOR-like1* gene significantly inhibited tomato fruit ripening. The *NOR-like1* homozygous mutant obtained by CRISPR technology exhibited a 14-day delay in the breaker stage compared to WT, and the mutant fruit did not turn red completely (Figure 4.4).

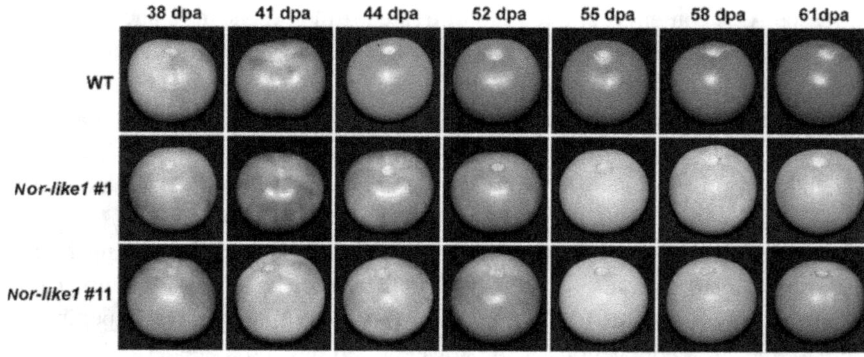

Figure 4.4 Fruit phenotype of the *nor-like1* mutant (Gao et al. 2018/with permission from Oxford University Press) dpa: days postanthesis.

Tests for physiological parameters showed that compared to WT, the *nor-like1* mutant exhibited obviously reduced ethylene production in the fruit, repressed chlorophyll degradation and carotenoid accumulation after the breaker stage, and increased fruit firmness (Gao et al. 2018).

RNA-Seq analysis showed that the mutation of *NOR-like1* gene affected the expression of key genes in ethylene biosynthesis and signal transduction pathway, carotenoid biosynthesis, and fruit softening as well as the expression of key TFs that regulate fruit ripening. These findings indicate that NOR-like1 regulates fruit-ripening-related gene expression to control tomato fruit ripening. Gene expression analysis, promoter analysis, electrophoretic mobility shift assay (EMSA), chromatin immunoprecipitation (ChIP), and dual luciferase reporter assay (DLR) showed that NOR-like1 could regulate ethylene biosynthesis by directly binding to the promoters of *SlACS2* and *SlACS4*. It could also directly bind to the promoters of *SlSGR1* and *SlGgpps2* to regulate chlorophyll degradation and carotenoid accumulation, and it directly binds to the promoters of *SlPG2a*, *SlPL*, *SlCEL2*, and *SlEXP1* to regulate tomato fruit softening. The mechanism of NOR-like1 regulating fruit ripening is as shown in Figure 4.5 (Gao et al. 2018).

Recent studies have found that in the *nor* mutants obtained by CRISPR/Cas9 technology, fruit ripening was partially inhibited; the fruit was an orange color, which was different from the *nor* natural mutants in which fruit ripening was completely blocked and that exhibited a yellow fruit phenotype (Gao et al. 2019; Wang et al. 2019), as shown in Figure 4.6. Moreover, CRISPR technology was used to mutate the remaining NOR gene in the *nor* natural mutant, which could partially restore the nonripening mutant phenotype (Wang et al. 2019). On the other hand, overexpression of NOR in the *nor* natural mutant, the unripe phenotype of the mutant could be recovered partially (Gao et al. 2020), as shown in in

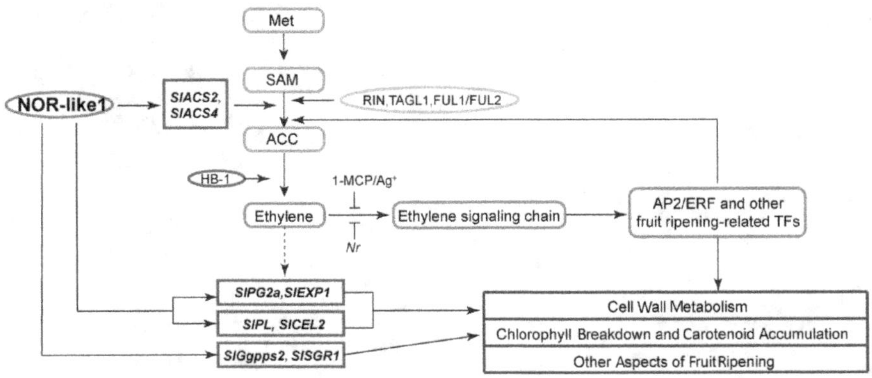

Figure 4.5 Working model of the regulatory mechanism of NOR-like1 in tomato fruit ripening (Gao et al. 2018/Oxford University Press/CC BY 4.0.).

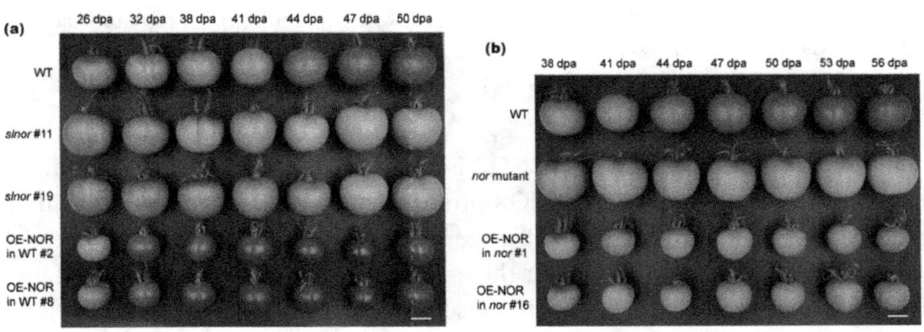

Figure 4.6 The phenotypes of CR-NOR and OE-NOR fruits in the WT background (a) and the phenotype of OE-NOR fruits in the *nor* background (b) (Gao et al. 2020/with permission from Oxford University Press).

Figure 4.6b. The above studies indicated that the *nor* natural mutant was not a null mutant but a gain-of-function mutant. NAC-NOR protein can enter the nucleus, bind to and activate the promoters of its target genes, while NOR186 also can enter the nucleus and bind to the promoter region of the target genes but cannot activate their promoters and play the role of a dominant-negative protein (Figure 4.7).

Despite the function of the *nor* natural mutants was reassessed, the fact is that NOR is a key transcription factor regulating tomato fruit ripening, involving in the regulation of ethylene biosynthesis, carotenoid accumulation, fruit softening and flavor volatiles metabolism of tomato fruit (Gao et al. 2020, 2022). In addition, NOR could bind to and activate the promoter of *SlDML2* (DNA demethylase), affect the promoter methylation level of flavor volatiles biosynthesis genes, regulate tomato fruit flavor volatiles accumulation via interacting with SlDML2 (Gao et al. 2022).

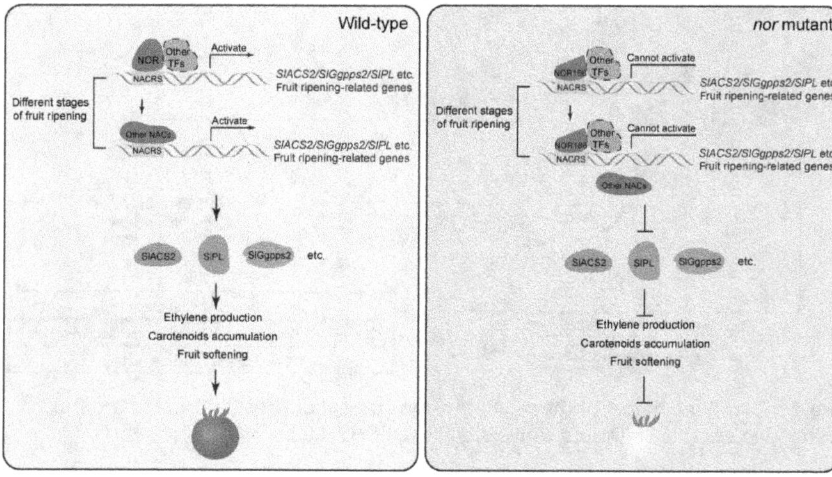

Figure 4.7 Model of NAC-NOR and the mutant NOR186 in regulation of tomato fruit ripening (Gao et al. 2020/Oxford University Press/CC BY 4.0.).

Recent studies have shown that NOR is subject to posttranslational modifications (PTMs) of proteins. Met in NOR protein is susceptible to oxidation, which affects the function of proteins. Sulfoxide reductase E4 and SlMsrB2 can directly interact with NOR, partially repair the oxidation of Met in NOR, so as to stabilize the DNA-binding ability and transcriptional regulation activity of NOR (Jiang et al. 2020).

Jian et al. (2021) found that SlNAC6 induced the early expression of key genes SlACS2, SlACS4, and SlACO1 in ethylene biosynthesis. It also activates the expression of SlNCED1, SlNCED2, SlABA2, SlAO1, and SlAO2 in the ABA biosynthesis pathway, thus promoting the accumulation of ethylene and ABA, and positively regulating fruit ripening.

Other transcription factors of NAC family in tomato, SlNAM1/2/3, also participate in tomato fruit ripening. As positive regulators of tomato fruit ripening, they could directly bind to and activate the promoters of *SlACS2* and *SlACS4*, as a result, regulate ethylene production (Gao et al. 2021; Lin et al. 2022). Furthermore, SlNAM2 and SlNAM3 are targeted by miR164a, directly interact with the promoters of genes involved in chloroplast development/photosynthesis, and activate their transcription, including *SlGLK1*, *SlCAB11*, *SlCAO*, *SlPOR3*, and *SlPsbP3* (Lin et al. 2022). They also bind to the promoters of other fruit-ripening-related genes and positively regulate their transcription, such as *SlACO1*, *SlE8*, *SlZDS*, *SlPSY1*, and *SlXTH3* (Lin et al. 2022).

4.3.2 *CpNAC1/3* Is Involved in Regulating the Ripening of Papaya

Although many genes in the carotenoid biosynthetic pathway in papaya have been identified (such as *CpPDS*, *CpZDS*, *CpLCY-Bs*, and *CpCYC-b*), the transcriptional regulatory network of these genes is still unclear. After propylene treatment, the key genes in the carotenoid biosynthetic pathway, phytoene desaturase genes (*CpPDSs*), and *CpNAC1*, were upregulated, consistent with the increased carotenoid content in papaya. EMSA and transient expression analysis have shown that the CpNAC1 protein directly binds to the NACBS in the promoter region of *CpPDS2/4* and activates its transcription. These results indicate that during papaya fruit ripening, CpNAC1 may activate the transcription of *CpPDSs* (such as *CpPDS2/4*) to positively regulate carotenoid biosynthesis in papaya (Fu et al. 2016).

CpNAC3 and CpMADS4 bind to and activate the promoters of *CpERF9* and *CpEIL5*, regulating ethylene signal transduction. Moreover, CpNAC3 can interact with CpMADS4 to enhance their transcriptional activation. These results suggest that CpNAC3 and CpMADS4 can independently or co-interaction to activate the ethylene signaling transduction pathway and positively regulate fruit ripening in papaya (Fu et al. 2021).

4.3.3 NAC Family TFs Are Involved in Fruit Deastringency in Persimmon

In 2015, Min et al. treated persimmon fruits with high concentrations of CO_2 (95%) and reduced the content of soluble tannin and increased the accumulation of acetaldehyde, thereby removing the astringency of persimmon fruit. Using RNA-Seq and rapid amplification of cDNA ends (RACE), they found six *DkNAC* genes that may be related to persimmon astringency removal. Transcriptional analysis showed that these six NAC genes responded to CO_2 differently: the expression of *DkNAC1*, *DkNAC3*, *DkNAC5*, and *DkNAC6* was significantly upregulated during persimmon fruit astringency removal. This finding indicates that NAC TFs are closely associated with fruit astringency removal (Min et al. 2015). The expression level of *DkNAC7/13/16* is also consistent with the fruit deastringency under high CO_2 concentration. After CO_2 treatment, DkNAC7 can regulate the process of postharvest persimmon fruit astringency removal by trans-activating the fruit astringency regulator *DkERF9* and pyruvate decarboxylase 2 (*DkPDC2*). DkNAC13/16 can respectively activate deastringency-related genes *DkERF9* and alcohol dehydrogenase 1 (*DkADH1*) to promote the deastringency process of persimmon fruit (Figure 4.8) (Jin et al. 2018; Jamil et al. 2019).

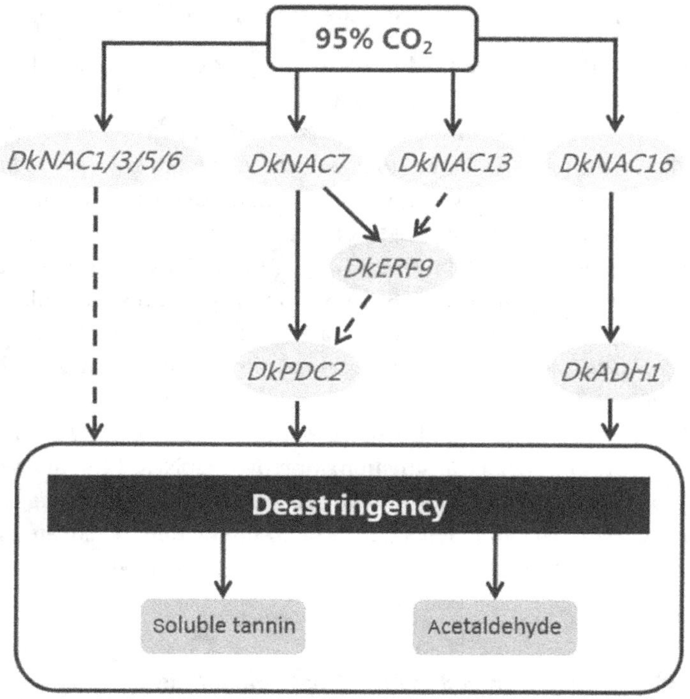

Figure 4.8 Synergistic reaction model of DkNACs on AHCA treatment.

4.3.4 NAC Family Transcription Factors Regulate Peach Fruit Ripening

In 2015, Zhou et al. analyzed the anthocyanin content in the blood-fleshed peach cultivar Dahongpao (DHP) and the white-fleshed cultivar Baifeng (BF). They found that the predominant component of anthocyanin was cyanidin-3-glucoside, and the anthocyanin content in the blood-fleshed peach was far higher than that in the white-fleshed peach. They further analyzed the expression levels of key genes in the anthocyanin biosynthetic pathway (*PAL*, *C4H*, *4CL*, *CHS*, *CHI*, *F3H*, *F3OH*, *DFR*, *LDOX*, and *UFGT*) in DHP and BF. These genes exhibited high expression levels in fruit at late developmental stages in DHP, but their expression was not detected, or was detected at very low levels, at early developmental stages in DHP fruit or late developmental stages in BF fruit. These results indicate that anthocyanin biosynthesis is regulated at the transcriptional level. Investigation of genes regulating anthocyanin biosynthesis revealed that the expression level of *PpMYB10.1* was closely related to anthocyanin synthesis in the fruit, indicating that the activation of *PpMYB10.1* is a critical step in anthocyanin biosynthesis. (Zhou et al , 2015).

In subsequent genetic mapping, the blood-flesh trait was mapped to a 200-kb interval on peach linkage group (LG) 5. Within this interval, a NAC family TF gene was found and designated as *BLOOD* (*BL*). However, overexpressing the *BL* gene did not affect the expression of *PpMYB10.1*. Since NAC family TFs usually function as homo- or heterodimers, Zhou et al. performed dual luciferase assays and a yeast two-hybrid experiment and showed that BL and PpNAC1 formed a heterodimer to activate the expression of *PpMYB10.1*. The absence of anthocyanin accumulation at the early stage of fruit ripening may be due to the inhibition of *PpNAC1* expression by PpSPL1.

PpNAC1 can also promote volatile esters biosynthesis during fruit ripening by directly binding to the promoter and activating the alcohol acyltransferase gene (*PpAAT1*). This pathway is located downstream of ethylene signal and remains conservative in tomato (SlNOR-SlAAT1) and apple fruit (MdNAC5-MdAAT1) (Cao et al. 2021).

In addition to BL and PpNAC1, PpNAP1, PpNAP4, and PpNAP6 may also involve in peach ripening regulation by regulating ethylene biosynthesis and responding to ABA signals (Li et al. 2016a). Prupe.4G187100, a homolog of tomato SlNOR, could control the time and process of fruit ripening in a conservative manner depending on ethylene (Tan et al. 2021). Guo et al. (2021) identified *PpNAC.A59* and *PpERF.A16* through RNA-Seq, which were highly expressed during peach fruit maturation. *PpNAC.A59* could directly bind to and activate the promoter of *PpERF.A16*. Moreover, *PpERF.A16* can further activate the expression of the key genes for ethylene biosynthesis, *PpACS1* and *PpACO1*, as well as the key gene for cell wall degradation, *PpendoPGM*, forming the cascade transduction of ripening signals and positively regulating fruit ripening.

4.3.5 NAC Family Transcription Factors Regulate the Ripening of Kiwifruit

In 2015, Nieuwenhuizen et al. conducted a comparative study in two kiwifruit (*Actinidia*) cultivars, *Actinidia arguta* and *Actinidia chinensis*, to study the biosynthetic mechanism of terpenes in kiwifruit. They found that in *Actinidia arguta*, terpene synthase AaTPS1 and DXS exhibited high gene expression levels and protein levels. Transient overexpression of *DXS* in *Nicotiana benthamiana* leaves elevated monoterpene synthesis mediated by AaTPS1 by more than 100-fold, indicating that DXS may regulate a key step of 2-C-methyl-d-erythritol 4-phosphate substrate flux in kiwifruit. Promoter analysis identified binding sites for NAC and ETHYLENE-INSENSITIVE3-Like (EIL) TFs in the *AaTPS1* promoter region. EMSA confirmed that AaNAC2, AaNAC3, and AaNAC4 can bind a 28-bp fragment of the proximal NAC-binding site in the *AaTPS1* promoter. The results above showed that NAC family TFs can regulate terpene synthesis in kiwifruit during fruit ripening (Nieuwenhuizen et al. 2015).

Exogenous methyl jasmonate (MeJA) can synergistically enhance the induction of exogenous ethylene on kiwifruit ripening, *AdNAC2/3* is induced, and AdNAC2/3 could bind to and activate the promoter of *AdACS1* to promote endogenous ethylene production (Wu et al. 2020). In addition to AdNAC2/3, AcNAC1/4/6/7 can also bind to and activate the promoter of *AcACS1*. AcNAC2-4 is responsible for regulating ethylene biosynthesis during ripening initiation (system I) and fruit ripening (system II), playing a positive role in kiwifruit ripening (Nieuwenhuizen et al. 2021). AdNAC6/7 broadly regulate metabolic processes during kiwifruit fruit ripening, they could activate the expression of the cell wall degradation-related gene *AdMAN1* to promote fruit softening. AdNAC6/7 can also bind to and activate the promoter of *AaTPS1* to promote the synthesis of terpene aroma compounds (Wang et al. 2020). It can also form homodimers or heterodimers in the nucleus. Their transcripts can be targeted for degradation by miR164; the pathway is located downstream of ethylene signaling and is repressed by ethylene (Figure 4.9) (Wang et al. 2020).

4.3.6 *EjNAC1* Regulates Lignin Biosynthesis in Loquat Fruit

Lignin is an important component of the secondary cell wall; its content greatly impacts the utilization of plant products. Loquat is a typical fruit that accumulates lignin during postharvest low-temperature storage. In 2015, Xu et al. cloned two NAC family TFs and named them *EjNAC1* and *EjNAC2*. They found that these

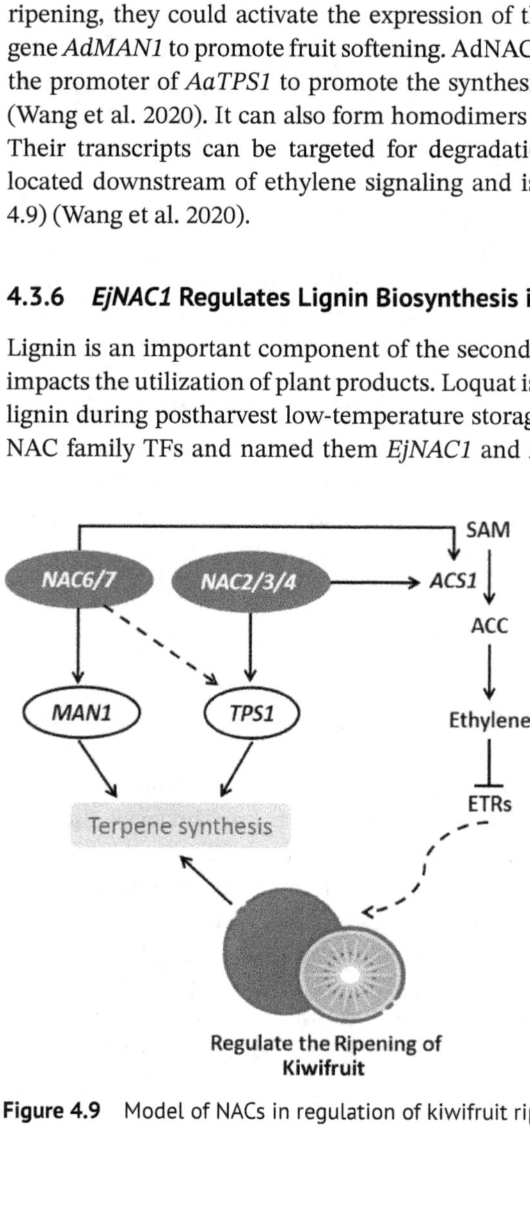

Figure 4.9 Model of NACs in regulation of kiwifruit ripening.

two TFs both exhibited transcriptional activity in yeast and that their expression in lignin-rich stem was significantly higher than that in the pulp. However, only the expression of *EjNAC1* was induced by low temperature and repressed by heat, which can alleviate lignification, whereas the expression of *EjNAC2* was not correlated with fruit lignification. Further study showed that *EjNAC1* could activate the promoters of genes related to the lignin biosynthetic pathway in loquat and *Arabidopsis*. Transient overexpression of *EjNAC1* in tobacco leaf led to lignin accumulation and the induction of expression of endogenous lignin biosynthetic genes. These results showed that EjNAC1 is a positive regulator of lignin biosynthesis in loquat fruit (Xu et al. 2015). The transcription level of *EjNAC3* is also consistent with the degree of fruit lignification. It can physically combine with the promoter of *EjCAD-like* gene and trans activate its transcription, indicating that EjNAC3 is an activator of loquat lignification (Ge et al. 2017).

4.3.7 NAC Family TFs Are Involved in Climacteric Fruit Ripening in *Pyrus Ussuriensis*

Chinese pear (*Pyrus ussuriensis*) is a typical climacteric fruit that produces large amounts of ethylene during fruit ripening. Although researchers have studied the molecular mechanism of climacteric fruit ripening in depth, the current knowledge is still not comprehensive. As a result, Huang et al. used RNA-Seq to comparatively analyze the transcriptomes of pre- and postclimacteric fruits of Chinese pear and discovered 3279 differentially expressed genes (DEGs). Gene Ontology (GO) analysis found that these DEGs were enriched in biological process, molecular function, and cellular components. DEGs included genes related to plant hormones, such as ethylene, abscisic acid, auxin, gibberellin, and brassinosteroids. There were also TF-related genes, such as *MADS*, *NAC*, *WRKY*, and *HSF*. In addition, genes encoding DNA methylation-related methyltransferase, cytoskeletal proteins, and heat-shock proteins (HSP) exhibited differential expression between pre- and postclimacteric fruit ripening. RT-PCR results were consistent with RNA-Seq. This study confirmed that NAC family TFs are involved in climacteric fruit ripening in Chinese pear at the transcriptional level (Huang et al. 2014).

4.3.8 The Function of NAC Family TFs in Banana Fruit Ripening

In 2012, Shan et al. cloned six NAC TF genes from banana fruit and named them *MaNAC1–MaNAC6*. Subcellular localization determined that MaNAC1–MaNAC5 were localized in the nucleus, while MaNAC6 was distributed throughout the entire cell. A transactivation assay in yeast showed that MaNAC4 and MaNAC6, as well as their C-terminal regions, possessed transactivation activity. Analyzing the expression pattern of *MaNACs* in four treatments including the control and treatments with ethylene, 1-MCP, and ethylene with 1-MCP, during postharvest ripening showed that these genes exhibited different expression patterns in peel and pulp. Upon ethylene treatment, the expression of *MaNAC1* and *MaNAC6* was upregulated in peel and pulp, consistent with the increased level of ethylene. In contrast, *MaNAC3* in the peel and pulp and *MaNAC5* in the peel did not exhibit any significant changes in their expression levels; *MaNAC4* in the peel and pulp and *MaNAC6* in the peel had decreased expression, while the expression levels of *MaNAC5* and *MaNAC6* in the peel were slightly upregulated. Furthermore, ethylene treatment could activate the *MaNAC2* promoter and increase the expression of *MaNAC2* during fruit ripening. More importantly, yeast two-hybrid and bimolecular fluorescence complementation analyses confirmed that MaNAC1/MaNAC2 proteins could interact with MaEIN5, which was downstream of ethylene signal transduction. The results from these experiments indicate that the banana NAC family TFs, such as MaNAC1/MaNAC2, may regulate banana fruit ripening through interacting with genes in the ethylene signal transduction pathway (Shan et al. 2012).

Later, Shan performed a more in-depth study of the regulatory mechanism of NAC family TFs involved in banana fruit ripening and found that MaNAC1/2 and RING-type E3 ubiquitin ligase MaXB3 function antagonistically to affect ethylene signaling and biosynthesis. Transcriptional activity analysis showed that MaNAC1 and MaNAC2 are transcriptional repressors. Gel retardation analysis, immunoprecipitation, and transient expression analysis showed that MaNAC1 and MaNAC2 could directly bind to the promoters of *MaRTH1* and *MaERF11*, which were negative regulators of ethylene signaling, to inhibit their expression. Protein interaction and ubiquitin degradation experiments showed that the RING-type E3 ubiquitin ligase MaXB3 interacts with MaNAC2, MaACS1, and MaACO1 and mediates the ubiquitination of MaNAC2, MaACS1, and MaACO1 for 26S proteasome degradation, thereby suppressing the ability of MaNAC2 to act as transcriptional repressor as well as ethylene biosynthesis. More interestingly, MaNAC1 and MaNAC2 could directly bind to the promoter of *MaXB3* and inhibit its expression, suggesting a feedback regulatory mechanism that inhibited the MaXB3-mediated protein degradation by ubiquitination and maintained the stability of MaNAC2, MaACS1, and MaACO1 proteins (Shan 2016).

To further investigate other banana NAC family TFs that have different mechanisms from that of MaNAC1/2, Shan isolated another 11 NAC family members from the banana genome and named them *MaNAC7–MaNAC17*. Through expression and transcription activity analyses, *MaNAC10* and *MaNAC11* were found to be significantly upregulated during banana fruit ripening, while *MaNAC12*, *MaNAC13*, and *MaNAC14* were significantly downregulated. The results showed that MaNAC10 and MaNAC11 synergistically activated the expression of *MaACS1* and *MaACO1* directly, while MaNAC12 repressed the expression of *MaACS1* and *MaACO1*. Regulatory factors upstream of MaNAC10–MaNAC14 were further screened. MaSPL1 was found to bind to the promoter of *MaNAC10/11* and activate its expression, while MaMADS1 could bind to the promoter of *MaNAC12* to repress its expression. Additionally, the MaXB3 protein could interact with MaNAC10 and MaNAC11 and degrade these proteins through ubiquitination, thereby suppressing the ability of MaNAC10/11 to transcriptionally activate *MaACS1* and *MaACO1*. MaSPL1, MaMADS1, and MaNAC10–12 formed a cascade network to regulate fruit ripening in banana (Shan 2016), as shown in Figure 4.10.

Yan et al. (2021) found that under oxidative stress, the expression levels of *MaNAC42* and *MaMsrB2* in banana fruits increased significantly. Further studies showed that MaNAC42 could directly bind to promoters of a series of genes related to oxidative stress and fruit ripening, such as dehydration responsive element binding 1 (*MaDREB1*) and *MaERF113*. Moreover, the sulfonation of MaNAC42 reduces its ability to bind and activate downstream target genes. MaMsrB2 can interact with MaNAC42 to repair the oxidative damage of MaNAC42, thus restoring the DNA-binding ability of MaNAC42.

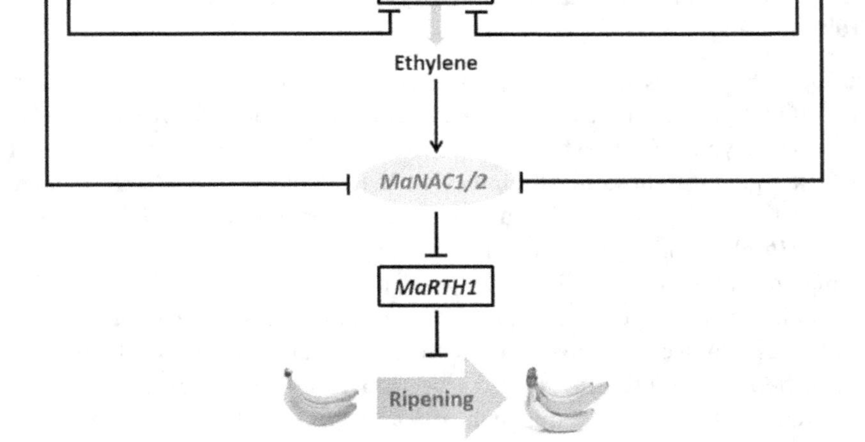

Figure 4.10 Working model of NAC TFs in banana fruit ripening (Peter Zijlstra/Shutterstock).

4.3.9 FcNAC1 and FaRIF Are Involved in Pectin Metabolism during Ripening of Strawberry Fruit

Carrasco-Orellana et al. isolated the *FcNAC1* gene from strawberry (*Fragaria chiloensis*). The expression of *FcNAC1* was upregulated during ripening of strawberry fruit. Dual luciferase assay of transiently transformed *Nicotiana benthamiana* leaves showed that FcNAC1 could activate the promoter of *FcPL*, indicating that FcNAC1 is involved in pectin metabolism during strawberry fruit ripening (Carrasco-Orellana et al. 2018). Another transcription factor of NAC family, FarIF, is highly expressed in strawberry and regulates a series of physiological changes during fruit ripening. Silencing of FaRIF gene by RNAi technology significantly delayed the ripening process of strawberry fruit, resulted in a lighter red color, reduced anthocyanin and sugar content, and firmer strawberry fruit (Martin-Pizarro et al. 2021).

4.3.10 The Interaction between LcNAC13 and LcR1MYB1 Regulates Anthocyanin Biosynthesis in Litchi Fruit

Jiang et al. isolated a NAC family TF gene from litchi, *LcNAC13*, whose expression level was upregulated as litchi fruit ripened. EMSA and transient expression experiments showed that LcNAC13 negatively regulated the expression of anthocyanin biosynthesis-related genes *LcCHS1/2*, *LcCHI*, *LcF3H*, *LcF3H*, *LcDFR*, and *LcMYB1*. In addition, LcNAC13 and LcR1MYB1 interact, and together, they regulate anthocyanin biosynthesis during fruit ripening in litchi (Jiang et al. 2019).

4.3.11 The Role of NAC Family Transcription Factors in Apple Fruit Ripening

In apple, both MdNAC1 and MdNAC2 could interact with MdRTE1a/b, an important regulator of ethylene signaling. *MdNAC1* and *MdRTE1a* are mainly expressed in young fruits, whereas *MdNAC2* and *MdRTE1b* are highly expressed in fruit ripening. These results indicated that *MdNAC1* and *MdRTE1a* may involve in the regulation of young fruit development, while *MdNAC2* and *MdRTE1b* play synergistic roles in fruit ripening by regulating ethylene signaling (Zhang et al. 2018). Zhang et al. identified 13 *NAC* genes that are differentially expressed during apple fruit development and ripening. During postharvest storage, ethylene and 1-MCP could induce or inhibit the expression of *MdNAC1a* and *MdNAC78*, respectively. MdNAC18.1 is a homologue of SlNOR in tomato. Overexpression of *MdNAC18*.1 in the *nor* mutant could enhance the expression of *SlACS2* and partially restore the unripe phenotype of *nor* (Migicovsky et al. 2021).

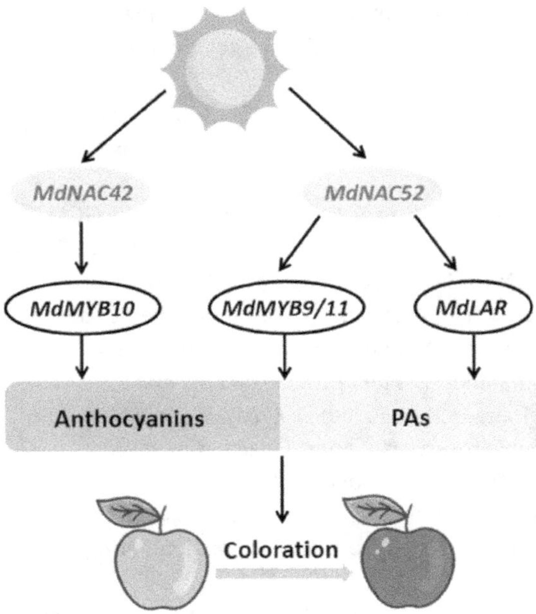

Figure 4.11 Proposed model for MdNAC52-regulated anthocyanin and PA biosynthesis.

The transcript levels of *MdNAC42*/52 in apple are increased during apple fruit coloration. Further research showed that MdNAC42 could interact with *MdMYB10*, an important positive regulator of anthocyanin biosynthesis, to promote the accumulation of anthocyanins during apple fruit ripening (Zhang et al. 2020). MdNAC52 could bind to and activate the promoters of proanthocyanidins (PA), positive regulators (*MdMYB9* and *MdMYB11*), and the key enzyme (leucocyanin reductase, *LAR*) in anthocyanin biosynthesis pathway, promote the accumulation of PA and anthocyanins during fruit ripening (Figure 4.11) (Sun et al. 2019).

References

Aida, M., Ishida, T., Fukaki, H. et al. (1997). Genes involved in organ separation in Arabidopsis: an analysis of the cup-shaped cotyledon mutant. *Plant Cell* 9: 841–857.

Baker, C.C., Sieber, P., Wellmer, F., and Meyerowitz, E.M. (2005). The early extra petals1 mutant uncovers a role for microRNA miR164c in regulating petal number in Arabidopsis. *Curr Biol* 15: 303–315.

Bhattacharjee, P., Das, R., Mandal, A., and Kundu, P. (2017). Functional characterization of tomato membrane-bound NAC transcription factors. *Plant Mol Biol* 93: 511–532.

Cao, X., Wei, C., Duan, W. et al. (2021). Transcriptional and epigenetic analysis reveals that NAC transcription factors regulate fruit flavor ester biosynthesis. *Plant J* 106: 785–800.

Carrasco-Orellana, C., Stappung, Y., Mendez-Yanez, A. et al. (2018). Characterization of a ripening-related transcription factor FcNAC1 from Fragaria chiloensis fruit. *Sci Rep* 8: 10524.

Cohen, S.X., Moulin, M., Hashemolhosseini, S. et al. (2003). Structure of the GCM domain-DNA complex: a DNA-binding domain with a novel fold and mode of target site recognition. *EMBO J* 22: 1835–1845.

De Clercq, I., Vermeirssen, V., Van Aken, O. et al. (2013). The Membrane-Bound NAC transcription factor ANAC013 functions in mitochondrial retrograde regulation of the oxidative stress response in arabidopsis. *Plant Cell* 25: 3472–3490.

de Oliveira, T.M., Cidade, L.C., Gesteira, A.S. et al. (2011). Analysis of the NAC transcription factor gene family in citrus reveals a novel member involved in multiple abiotic stress responses. *Tree Genet Genomes* 7: 1123–1134.

Duan, M., Zhang, R., Zhu, F. et al. (2017). A lipid-anchored NAC transcription factor is translocated into the nucleus and activates glyoxalase I expression during drought stress. *Plant Cell* 29: 1748.

Ernst, H.A., Olsen, A.N., Larsen, S., and Lo, L.L. (2004). Structure of the conserved domain of ANAC, a member of the NAC family of transcription factors. *Embo Rep* 5: 297–303.

Fu, C., Chen, H., Gao, H. et al. (2021). Papaya CpMADS4 and CpNAC3 co-operatively regulate ethylene signal genes CpERF9 and CpEIL5 during fruit ripening. *Postharvest Biol Technol* 175: 111485.

Fu, C., Han, Y., Fan, Z. et al. (2016). The papaya transcription factor CpNAC1 modulates carotenoid biosynthesis through activating phytoene desaturase GenesCpPDS2/4 during fruit ripening. *J Agr Food Chem* 64: 5454–5463.

Gao, Y., Fan, Z., Zhang, Q. et al. (2021). A tomato NAC transcription factor, SlNAM1, positively regulates ethylene biosynthesis and the onset of tomato fruit ripening. *Plant J* 108: 1317–1331.

Gao, Y., Lin, Y., Xu, M. et al. (2022). The role and interaction between transcription factor NAC-NOR and DNA demethylase SlDML2 in the biosynthesis of tomato fruit flavor volatiles. *New Phytol* 235: 1913–1926.

Gao, Y., Wei, W., Fan, Z. et al. (2020). Re-evaluation of the nor mutation and the role of the NAC-NOR transcription factor in tomato fruit ripening. *J Exp Bot* 71: 3560–3574.

Gao, Y., Wei, W., Zhao, X. et al. (2018). A NAC transcription factor, NOR-like1, is a new positive regulator of tomato fruit ripening. *Hortic Res-England* 5.

Gao, Y., Zhu, N., Zhu, X. et al. (2019). Diversity and redundancy of the ripening regulatory networks revealed by the fruitENCODE and the new CRISPR/Cas9 CNR and NOR mutants. *Hortic Res-England* 6.

Ge, H., Zhang, J., Zhang, Y. et al. (2017). EjNAC3 transcriptionally regulates chilling-induced lignification of loquat fruit via physical interaction with an atypical CAD-like gene. *J Exp Bot* 68: 5129–5136.

Giovannoni, J.J. (2007). Fruit ripening mutants yield insights into ripening control. *Curr Opin Plant Biol* 10: 283–289.

Greve, K., La Cour, T., Jensen, M.K. et al. (2003). Interactions between plant RING-H2 and plant-specific NAC (NAM/ATAF1/2/CUC2) proteins: RING-H2 molecular specificity and cellular localization. *Biochem J* 371: 97–108.

Guo, H.S., Xie, Q., Fei, J.F., and Chua, N.H. (2005). MicroRNA directs mRNA cleavage of the transcription factor NAC1 to downregulate auxin signals for arabidopsis lateral root development. *Plant Cell* 17: 1376–1386.

Guo, S., Dai, S., Singh, P.K. et al. (2018). A Membrane-Bound NAC-Like transcription factor OsNTL5 represses the flowering in Oryza sativa. *Front Plant Sci* 9: 555.

Guo, Z., Zhang, Y., Yao, J. et al. (2021). The NAM/ATAF1/2/CUC2 transcription factor PpNAC.A59 enhances PpERF.A16 expression to promote ethylene biosynthesis during peach fruit ripening. *Hortic Res* 8: 209.

He, X., Mu, R., Cao, W., et al. (2005). AtNAC2, a transcription factor downstream of ethylene and auxin signaling pathways, is involved in salt stress response and lateral root development. *Plant J* 44: 903–916.

Hegedus, D., Yu, M., Baldwin, D. et al. (2003). Molecular characterization of Brassica napus NAC domain transcriptional activators induced in response to biotic and abiotic stress. *Plant Mol Biol* 53: 383–397.

Hu, R., Qi, G., Kong, Y. et al. (2010). Comprehensive analysis of NAC domain transcription factor gene family in Populus trichocarpa. *BMC Plant Biol* 10: 145.

Huang, G., Li, T., Li, X. et al. (2014). Comparative transcriptome analysis of climacteric fruit of Chinese pear (Pyrus ussuriensis) reveals new insights into fruit ripening. *PLOS One* 9: e107562.

Jamil, W., Wu, W., Ahmad, M. et al. (2019). High-CO2/hypoxia-modulated NAC transcription factors involved in de-astringency of persimmon fruit. *Sci Hortic* 252: 201–207.

Jian, W., Zheng, Y., Yu, T. et al. (2021). SlNAC6, A NAC transcription factor, is involved in drought stress response and reproductive process in tomato. *J Plant Physiol* 264: 153483.

Jiang, G., Li, Z., Song, Y. et al. (2019). LcNAC13 physically interacts with LcR1MYB1 to Coregulate Anthocyanin Biosynthesis-Related Genes during Litchi fruit ripening. *Biomolecules* 9.

Jiang, G., Zeng, J., Li, Z. et al. (2020). Redox regulation of the NOR transcription factor is involved in the regulation of fruit ripening in tomato. *Plant Physiol* 183: 671–685.

Jin, J., Zhang, H., Kong, L. et al. (2014). PlantTFDB 3.0: a portal for the functional and evolutionary study of plant transcription factors. *Nucleic Acids Res* 42: D1182–D1187.

Jin, R., Zhu, Q., Shen, X. et al. (2018). DkNAC7, a novel high-CO2/hypoxiainduced NAC transcription factor, regulates persimmon fruit de-astringency. *PLOS One* 13: e0194326.

Kaneda, T., Taga, Y., Takai, R. et al. (2009). The transcription factor OsNAC4 is a key positive regulator of plant hypersensitive cell death. *EMBO J* 28: 926–936.

Kim, S., Lee, A., Yoon, H., and Park, C. (2008). A membrane-bound NAC transcription factor NTL8 regulates gibberellic acid-mediated salt signaling in Arabidopsis seed germination. *Plant J* 55: 77–88.

Kim, S.G., Kim, S.Y., and Park, C.M. (2007a). A membrane-associated NAC transcription factor regulates salt-responsive flowering via FLOWERING LOCUS T in Arabidopsis. *Planta* 226: 647–654.

Kim, S.Y., Kim, S.G., Kim, Y.S. et al. (2007b). Exploring membrane-associated NAC transcription factors in Arabidopsis: implications for membrane biology in genome regulation. *Nucleic Acids Res* 35: 203–213.

Kim, Y.S., Kim, S.G., Park, J.E. et al. (2006). A membrane-bound NAC transcription factor regulates cell division in Arabidopsis. *Plant Cell* 18: 3132–3144.

Kou, X., Liu, C., Han, L. et al. (2016). NAC transcription factors play an important role in ethylene biosynthesis, reception and signaling of tomato fruit ripening. *Mol Genet Genomics* 291: 1205–1217.

Laufs, P., Peaucelle, A., Morin, H., and Traas, J. (2004). MicroRNA regulation of the CUC genes is required for boundary size control in Arabidopsis meristems. *Development* 131: 4311–4322.

Li, F., Li, J., Qian, M. et al. (2016a). Identification of peach NAP transcription factor genes and characterization of their expression in vegetative and reproductive organs during development and senescence. *Front Plant Sci* 7: 147.

Li, P., Wind, J.J., Shi, X. et al. (2011a). Fructose sensitivity is suppressed in Arabidopsis by the transcription factor ANAC089 lacking the membrane-bound domain. *Proc Natl Acad Sci* 108: 3436–3441.

Li, S., Wang, N., Ji, D. et al. (2016b). Evolutionary and functional analysis of membrane-bound NAC transcription factor genes in soybean. *Plant Physiol* 172: 1804–1820.

Li, W., Han, L., Qian, Y.Q., and Sun, Z.Y. (2011b). Characteristics and functions of NAC transcription factors in plants. *Chin J Appl Environ Biol* 17: 596–606.

Lin, D., Zhu, X., Qi, B. et al. (2022). SlMIR164A regulates fruit ripening and quality by controlling SlNAM2 and SlNAM3 in tomato. *Plant Biotechnol J* 20: 1456–1469.

Ma, N., Feng, H., Meng, X. et al. (2014). Overexpression of tomato SlNAC1 transcription factor alters fruit pigmentation and softening. *BMC Plant Biol* 14: 351.

Ma, N.N. (2013). Functional of analysis of tomato NAC1 transcription factor. Shandong Agricultural University. 123.

Martín-Pizarro, C., Vallarino, J.G., Osorio, S., et al. (2021). The NAC transcription factor FaRIF controls fruit ripening in strawberry. *Plant Cell* 33: 1574–1593.

Meng, C., Yang, D., Ma, X. et al. (2016). Suppression of tomato SlNAC1 transcription factor delays fruit ripening. *J Plant Physiol* 193: 88–96.

Migicovsky, Z., Yeats, T., Watts, S. et al. (2021). Apple ripening is controlled by a NAC transcription factor. *Front Genet* 12: 671300.

Min, T., Wang, M., Wang, H. et al. (2015). Isolation and expression of NAC genes during persimmon fruit postharvest astringency removal. *Int J Mol Sci* 16: 1894–1906.

Nakashima, K., Takasaki, H., Mizoi, J. et al. (2012). NAC transcription factors in plant abiotic stress responses. *Biochim Biophys Acta* 1819: 97–103.

Ng, S., Ivanova, A., Duncan, O. et al. (2013). A membrane-bound NAC transcription factor, ANAC017, mediates mitochondrial retrograde signaling in Arabidopsis. *Plant Cell* 25: 3450–3471.

Nieuwenhuizen, N., Chen, X., Pellan, M. et al. (2021). Regulation of wound ethylene biosynthesis by NAC transcription factors in kiwifruit. *BMC Plant Biol* 21: 411.

Nieuwenhuizen, N.J., Chen, X., Wang, M.Y. et al. (2015). Natural variation in monoterpene synthesis in kiwifruit: transcriptional regulation of terpene synthases by NAC and ETHYLENE-INSENSITIVE3-like transcription factors. *Plant Physiol* 167: 1243–1258.

Nuruzzaman, M., Manimekalai, R., Sharoni, A.M. et al. (2010). Genome-wide analysis of NAC transcription factor family in rice. *Gene* 465: 30–44.

Olsen, A.N., Ernst, H.A., Lo Leggio, L., and Skriver, K. (2005a). NAC transcription factors: structurally distinct, functionally diverse. *Trends Plant Sci* 10: 79–87.

Olsen, A.N., Ernst, H.A., Lo Leggio, L. et al. (2004). Preliminary crystallographic analysis of the NAC domain of ANAC, a member of the plant-specific NAC transcription factor family. *Acta Crystallogr Sec D Biol Crystallogr* 60: 112–115.

Olsen, A.N., Ernst, H.A., Lo Leggio, L., and Skriver, K. (2005b). DNA-binding specificity and molecular functions of NAC transcription factors. *Plant Sci* 169: 785–797.

Ooka, H., Satoh, K., Doi, K. et al. (2003). Comprehensive analysis of NAC family genes in Oryza sativa and Arabidopsis thaliana. *DNA Res* 10: 239–247.

Peng, H., Yu, X.W., Cheng, H.Y. et al. (2010). A survey of functional studies of the plant-specific nac transcription factor family. *Bull Bot* 45: 236–248.

Rhoades, M.W., Reinhart, B.J., Lim, L.P. et al. (2002). Prediction of plant microRNA targets. *Cell* 110: 513–520.

Riechmann, J.L., Heard, J., Martin, G. et al. (2000). Arabidopsis transcription factors: genome-wide comparative analysis among eukaryotes. *Science* 290: 2105–2110.

Rushton, P.J., Bokowiec, M.T., Han, S. et al. (2008). Tobacco transcription factors: novel insights into transcriptional regulation in the Solanaceae. *Plant Physiol* 147: 280–295.

Seo, P.J., Kim, M.J., Song, J. et al. (2010). Proteolytic processing of anArabidopsis membrane-bound NAC transcription factor is triggered by cold-induced changes in membrane fluidity. *Biochem J* 427: 359–367.

Seo, P.J., Kim, S.G., and Park, C.M. (2008). Membrane-bound transcription factors in plants. *Trends Plant Sci* 13: 550–556.

Shan, W. (2016). Mechanism Analysis of NAC transcription factors in regulation of chilling tolerance and ripening in banana fruit. South China Agricultural University. 143.

Shan, W., Kuang, J., Chen, L. et al. (2012). Molecular characterization of banana NAC transcription factors and their interactions with ethylene signalling component EIL during fruit ripening. *J Exp Bot* 63: 5171–5187.

Souer, E., van Houwelingen, A., Kloos, D. et al. (1996). The no apical meristem gene of Petunia is required for pattern formation in embryos and flowers and is expressed at meristem and primordia boundaries. *Cell* 85: 159–170.

Sun, Q., Jiang, S., Zhang, T. et al. (2019). Apple NAC transcription factor MdNAC52 regulates biosynthesis of anthocyanin and proanthocyanidin through MdMYB9 and MdMYB11. *Plant Sci* 289: 110286.

Takada, S., Hibara, K., Ishida, T., and Tasaka, M. (2001). The CUP-SHAPED COTYLEDON1 gene of Arabidopsis regulates shoot apical meristem formation. *Development* 128: 1127–1135.

Tan, Q., Li, S., Zhang, Y. et al. (2021). Chromosome-level genome assemblies of five Prunus species and genome-wide association studies for key agronomic traits in peach. *Hortic Res* 8: 213.

Tang, Y., Zhao, C., Tan, S., and Xue, H. (2016). Arabidopsis type II Phosphatidylinositol 4-Kinase PI4Kγ5 regulates Auxin Biosynthesis and Leaf margin development through interacting with Membrane-Bound transcription factor ANAC078. *PLOS Genet* 12: e1006252.

Tran, L.S., Nakashima, K., Sakuma, Y. et al. (2007). Co-expression of the stress-inducible zinc finger homeodomain ZFHD1 and NAC transcription factors enhances expression of the ERD1 gene in Arabidopsis. *Plant J* 49: 46–63.

Tran, L.S.P. (2004). Isolation and functional analysis of Arabidopsis Stress-Inducible NAC transcription factors that bind to a drought-responsive cis-element in the early responsive to dehydration stress 1 promoter. *Plant Cell Online* 16: 2481–2498.

Wang, D., Yu, Y., Liu, Z. et al. (2016). Membrane-bound NAC transcription factors in maize and their contribution to the oxidative stress response. *Plant Sci* 250: 30–39.

Wang, R., Tavano, E.C.D.R., Lammers, M. et al. (2019). Re-evaluation of transcription factor function in tomato fruit development and ripening with CRISPR/Cas9-mutagenesis. *Sci Rep-UK* 9.

Wang, W., Wang, J., Wu, Y. et al. (2020). Genome-wide analysis of coding and non-coding RNA reveals a conserved miR164-NAC regulatory pathway for fruit ripening. *N Phytol* 225: 1618–1634.

Wu, Y., Liu, X., Fu, B. et al. (2020). Methyl jasmonate enhances ethylene synthesis in kiwifruit by inducing NAC genes that activate ACS1. *J Agric Food Chem* 68: 3267–3276.

Xie, Q., Guo, H.S., Dallman, G. et al. (2002). SINAT5 promotes ubiquitin-related degradation of NAC1 to attenuate auxin signals. *Nature* 419: 167–170.

Xu, Q., Wang, W., Zeng, J. et al. (2015). A NAC transcription factor, EjNAC1, affects lignification of loquat fruit by regulating lignin. *Postharvest Biol Tec* 102: 25–31.

Yamasaki, K., Kigawa, T., Inoue, M. et al. (2008). Structures and evolutionary origins of plant-specific transcription factor DNA-binding domains. *Plant Physiol Bioch* 46: 394–401.

Yan, H., Jiang, G., Wu, F. et al. (2021). Sulfoxidation regulation of transcription factor NAC42 influences its functions in relation to stress-induced fruit ripening in banana. *J Exp Bot* 72: 682–699.

Yang, S., Zhou, J., Watkins, C. et al. (2021). NAC transcription factors SNAC4 andSNAC9 synergistically regulate tomato fruit ripening by affecting expression of genes involved in ethylene and abscisic acidmetabolism and signal transduction. *Postharvest Biol Technol* 178: 111555.

Yang, X.N., Tian, Y., and Lu, X.Y. (2014). The regulation role of NAC transcription factors in plant growth and development. *Chem Bioeng* 31: 1–5.

Zhang, Q., Li, T., Zhang, L. et al. (2018). Expression analysis of NAC genes during the growth and ripening of apples. *Hortic Sci* 45: 1–10.

Zhang, S., Chen, Y., Zhao, L. et al. (2020). A novel NAC transcription factor, MdNAC42, regulates anthocyanin accumulation in red-fleshed apple by interacting with MdMYB10. *Tree Physiol* 40: 413–423.

Zhou, H., Lin-Wang, K., Wang, H. et al. (2015). Molecular genetics of blood-fleshed peach reveals activation of anthocyanin biosynthesis by NAC transcription factors. *Plant J* 82: 105–121.

Zhu, M., Chen, G., Zhang, J. et al. (2014b). The abiotic stress-responsive NAC-type transcription factor SlNAC4 regulates salt and drought tolerance and stress-related genes in tomato (*Solanum lycopersicum*). *Plant Cell Rep* 33: 1851–1863.

Zhu, M., Chen, G., Zhou, S. et al. (2014a). A new tomato NAC (NAM/ATAF1/2/CUC2) transcription factor, SlNAC4, functions as a positive regulator of fruit ripening and carotenoid accumulation. *Plant Cell Physiol* 55: 119–135.

5

Role of ERF Transcription Factors in Flesh Fruit Development and Ripening

Ethylene response factors (ERFs) are transcription factors (TFs) unique to plants that belong to the APETALA2/ERF (AP2/ERF) superfamily. ERFs play crucial roles in the ethylene signal transduction pathway and are important regulators of ethylene levels in plants. Studies have found that ERFs play key roles in plant growth and development as well as biotic and abiotic stresses in tomato, *Arabidopsis*, rice, and tobacco. Furthermore, ERFs have been shown to be involved in regulating fruit ripening in tomato, kiwifruit, banana, apple, papaya, sweet orange, and grape and are critical for fruit ripening. This chapter provides an overview of the plant ERF TFs and their biological functions.

5.1 Summary of the ERF Family

Ethylene, the first discovered gaseous hormone in plants, is involved in the regulation of various plant developmental processes. This hormone has important roles in agriculture. For example, it controls fruit ripening, organ senescence, and abscission. Ethylene also indirectly controls other aspects of plant growth and development, including seed germination, root formation, and floral development. The ethylene biosynthetic pathway has been thoroughly studied and includes three main steps, as follows.

1) The methionine cycle

First, methionine in plants is converted to S-adenosylmethionine (SAM) with the involvement of adenosine triphosphate (ATP). SAM is then converted to 1-aminocyclopropane-1-carboxylic acid (ACC) and 5′-methylthioadenosine (MTA). MTA is further hydrolyzed to 5-methylthioribose (MTR), which can be recycled to methionine through the methionine pathway. In ethylene biosynthesis, there is a cycle of methionine→SAM→MTA→methionine. In this cycle, the methylthio group can be recycled in plant tissues.

Transcriptional Regulation of Flesh Fruit Development and Ripening, First Edition. Yunbo Luo.

2) ACC synthesis

Because ACC is the direct precursor of ethylene biosynthesis, the process of converting SAM to ACC during ethylene biosynthesis is extremely important and is often considered the rate-limiting step in ethylene biosynthesis. The enzyme that catalyzes this process is ACC synthase (ACS). Therefore, ACS is the key enzyme that regulates ethylene biosynthesis. Aminoethoxyvinyl glycine (AVG) and aminooxyacetic acid (AOA) can inhibit the conversion of SAM to ACC.

3) Ethylene synthesis (ACC→ethylene)

ACC oxidase (ACO) is required to convert ACC to ethylene. This process requires oxygen. ACO is also a key enzyme for ethylene biosynthesis. Uncouplers and free radical scavengers can both inhibit the production of ethylene. The conversion of ACC to ethylene requires the cell membrane to be highly intact.

Ethylene biosynthesis occurs upstream of ethylene functions; its downstream biological functions are achieved via the ethylene signal transduction pathway. Because the ethylene signal transduction pathway is crucial to plant growth and development (Ju and Chang 2015), many studies have investigated this pathway. In the last few years, a considerable number of studies have focused on the mechanisms underlying the ethylene signal transduction pathway, and the most studied plant model is *Arabidopsis*. Major progress has been made in the understanding of the ethylene signal transduction pathway in *Arabidopsis*. A current model is as follows: ethylene → ETR family → CTR family → EIN3/EIL → ERF →expression of ethylene-related genes.

Among these factors, ERF TFs, as the downstream element in the ethylene signal transduction pathway, belong to the APETALA2/ERF (AP2/ERF) TF superfamily. ERF TFs play important roles in the ethylene signal transduction pathway and are crucial for plant growth and development. Downstream of the ethylene signal transduction pathway, there are five key members of the ethylene signal transduction system in *Arabidopsis*, including ETR1, ERS1, ETR2, ERS2, and EIN4. The expression of *ERFs*, as primary responsive genes, is first upregulated; then, the ERFs bind to the GCC box to upregulate the expression of secondary ethylene-responsive genes (Figure 5.1) (Ju and Chang 2015). A study has shown that in *Arabidopsis*, ethylene receptors are negative regulators; each dominant mutant exhibited reduced sensitivity to ethylene, while loss-of-function mutations in any single ethylene receptor had almost no effect in mutant plants (Qu et al. 2007).

The APETALA2/ERF TF superfamily is one of the largest TF families in plants. These TFs contain a 60–70 bp AP2/ERF domain and can be divided into three families: 1) the AP2 family, including two repetitive AP2/ERF domains; 2) the ERF family, containing one AP2/ERF domain; and 3) the RAV family, containing one AP2/ERF domain and one B3 domain (Nakano et al. 2006).

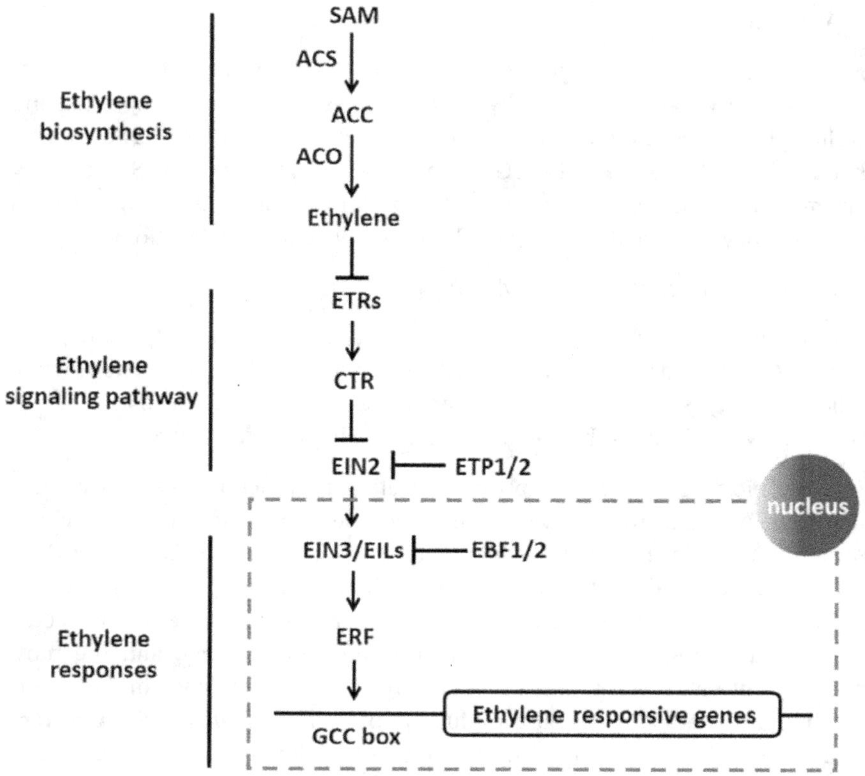

Figure 5.1 Ethylene biosynthesis and response in tomato.

1) AP2 family: In *Arabidopsis*, AP2 TFs mainly control three aspects of flowers: the establishment of the floral meristem, the development of floral organs, and the spatial and temporal expression of the flower homeotic genes. During early floral development, AP2 TFs mainly function in promoting the formation of floral meristem, together with APETALA1 (AP1), LEAFY (LFY), and CAULIFLOWER (CAL). The second function of AP2 is its regulation of floral organs. Evidence shows that loss-of-function *Arabidopsis* ap2 mutants exhibit weak growth, while gain-of-function ap2 mutants show good growth vigor. The third function of AP2 is that it can directly or indirectly regulate the expression of certain homeotic genes in the flower. These three functions of AP2 show that AP2 TFs play important roles in flower development in *Arabidopsis* (Jofuku et al. 1994).
2) ERF family: These TFs are members of the AP2/ERF superfamily. The ERF family is sometimes divided into two subfamilies: the ERF subfamily and the CBF/DREB subfamily. ERFs were first discovered as four DNA-binding proteins

in tobacco, i.e., ethylene response element-binding proteins 1, 2, 3, and 4 (EREBP1, 2, 3, and 4, also known as ERF1, 2, 3, and 4). These proteins can specifically bind to the GCC box, a DNA sequence involved in ethylene-responsive transcription (Ohme-Takagi and Shinshi 1995). After ERFs were discovered in tobacco, many ERF family proteins started to be understood. Scientists have found that ERFs possess various functions in physiological activities in the cell, such as affecting the signal transduction of many plant hormones and regulating metabolism, growth, and development.

3) In the RAV family, RAV1 and RAV2 were first identified as full-length cDNAs encoding proteins that contained a B3-like domain and an AP2/ERF domain in *Arabidopsis* (Nakano et al. 2006).

In the ERF family, the ERF subfamily contains alanine and aspartic acid as the 14th and 19th amino acids in the AP2/ERF domain, respectively, while the DREB subfamily contains valine and glutamate as the 14th and 19th amino acids, respectively (Sakuma et al. 2002). The AP2 domain was first discovered in the *Arabidopsis* AP2 protein that affects flower development. There are two repetitive AP2 domains in the AP2 protein: AP2-R1 (129–195 aa) and AP2-R2 (221–288 aa) (Jofuku et al. 1994).

AP2/ERF TFs are a large family broadly present in plants; many AP2/ERF TFs belong to the ERF family. Recently, a burst of research based on genome and transcriptome sequencing technologies has identified many AP2/ERF TFs which belong to the ERF family in different species, such as 122 in *Arabidopsis*, 139 in rice, 122 in grape, 91 in citrus, 155 in potato, 77 in tomato (Nakano et al. 2006; Zhuang et al. 2009; Licausi et al. 2010; Pirrello et al. 2012; Ito et al. 2014; Charfeddine et al. 2015).

ERF TFs often bind to DNA containing an 11-bp GCC box (TAAGAGCCGCC) in the 5′ promoter region and activate gene expression (Ohme-Takagi and Shinshi 1995). Studies in recent years have shown that ERFs could also bind to non-GCC box cis-elements. Tomato TF Pti4 belongs to the ERF family; transcriptomics and ChIP analysis showed that Pti4 could bind not only to the GCC box but also to non-GCC box cis-elements. A model of Pti4-mediated gene expression was then proposed (Figure 5.2) (Chakravarthy et al. 2003).

In addition, wound-induced proteins WRAF1 and WRAF2 also belong to the ERF family; they can bind to the vascular system-specific and wound-responsive cis-elements (VWRE; GAAAAGAAAATTTC) (Sasaki et al. 2007).

5.2 Biological Functions of the ERF Family

ERF TFs are widely present in plants and play important roles in plant growth and development, hormone signal transduction, as well as biotic and abiotic stresses. The rice FRIZZY PANICLE (FZP) protein belongs to the ERF family and plays

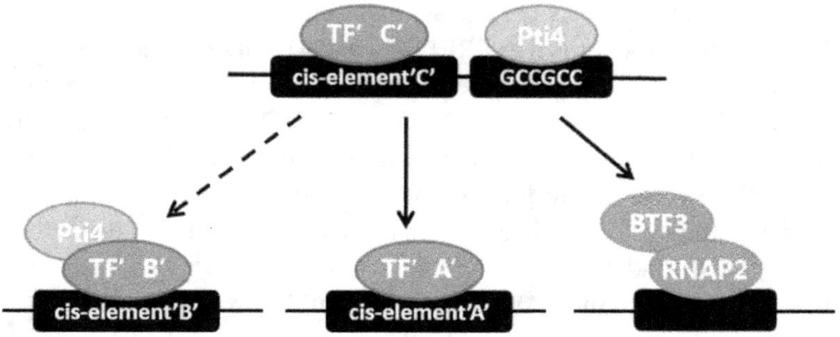

Figure 5.2 Models for Pti4-mediated gene expression.

important roles in rice spikelet development. In *fzp* mutants, the formation of florets is replaced by sequential rounds of branching (Komatsu et al. 2003). *Arabidopsis* LEAFY PETIOLE (LEP) protein, an ERF family member, is possibly involved in cell division in the marginal meristem to affect petiole development in *Arabidopsis* (van der Graaff et al. 2000). OsERF1, an ERF family member in rice, exhibits consistent expression levels across different tissues, and its expression can be upregulated by the plant hormone ethylene. Overexpression of *OsERF1* in *Arabidopsis* resulted in the upregulation of two ethylene-responsive genes: *PDF1.2* and *b-chitinase*; dwarf plants were observed, and seeds failed to develop and then they found that it also affected root growth in seedlings, while this effect can be restored by light (Hu et al. 2008, 2011). In *Arabidopsis*, overexpression of *ERF BUD ENHANCER* (*EBE*) promoted cell proliferation as well as axillary bud outgrowth and shoot branching, and the opposite phenotype was observed in lines where mRNA accumulation was inhibited by RNAi (Mehrnia et al. 2013). In tomato, *SlERF2* exists in two forms, spliced and unspliced (Figure 5.3). The functional *SlERF2* may arise from the transcript of the spliced form because the abundance of spliced transcripts in different tissues was far higher than that of the unspliced transcript. Overexpressing *SlERF2* in tomato led to premature seed germination significantly earlier than the wild-type seeds as well as enhanced hook formation of dark-grown seedlings during seed germination (Pirrello et al. 2006). In addition, *Sl-ERF.B3* integrates ethylene and auxin signaling via regulation of the expression of the auxin signaling component Sl-IAA27 (Liu et al. 2018). *TaERF8-2B* overexpression transgenic lines exhibited significantly reduced plant height and increased yield in wheat (Zhang et al. 2020). In apple, the *MdERF38* interacted with *MdMYB1*, which promoted anthocyanin biosynthesis in response drought stress (An et al. 2020).

ABR1 is an inhibitor of ABA and belongs to the ERF family. It plays important roles in ABA signal transduction. The *ABR1* mutation resulted in ABA hypersensitivity in seed germination and root growth in *Arabidopsis* (Pandey et al. 2005). Tomato stress-responsive factor (TSRF1) is an ERF family TF in tomato isolated

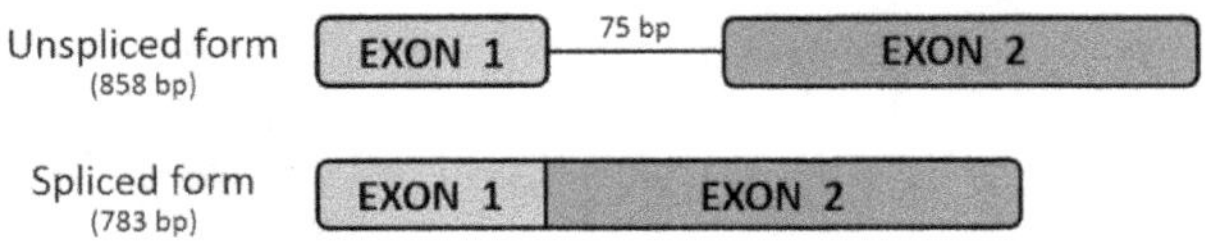

Figure 5.3 Gene structure of tomato *SlERF2*.

with a yeast one-hybrid assay; its expression can be upregulated by ethylene, salicylic acid, and *Ralstonia solanacearum* strain BJ1057 infection. TSRF1 can bind to the GCC box; overexpressing *TSRF1* in tomato and tobacco resulted in enhanced disease resistance against the *Ralstonia solanacearum* strain BJ1057 (Zhang et al. 2004b). Further study showed that TSRF1 can bind to a GCC box-like sequence, the CE1/GCC sequence, and affect ABA signal transduction. Overexpressing *TSRF1* in tobacco led to increased ABA sensitivity in seed germination, cotyledon expansion, and root elongation. In addition, TSRF1 can induce the expression of senescence-associated genes and accelerate the senescence of tobacco seedlings (Zhang et al. 2008). *Arabidopsis* ORA59 TF belongs to the ERF family and is a key integrator of jasmonic acid (JA) and the ethylene signaling pathway. Transgenic *Arabidopsis* with *ORA59* overexpression exhibited enhanced resistance against the fungus *Botrytis cinerea*, while transgenic *Arabidopsis* with silenced *ORA59* was more susceptible to *Botrytis cinerea* infection (Pre et al. 2008). Many studies have found that ERF family TFs are involved in plant biotic and abiotic stress responses. The expression of *Arabidopsis ERF1* significantly increased under salt and drought stress, and it is a positive regulator of biotic and abiotic stresses including salt, drought, and heat. When infected by pathogens, ERF1 binds to the GCC box, while under abiotic stress, it binds to the DRE element (G/ACCGAC) (Figure 5.4) (Cheng et al. 2013). *Arabidopsis* ERF1 also has significance roles in floral initiation. The floral initiation was delayed when *ERF1* obviously was overexpressing, while knockdown or knockout of *ERF1* was accelerated (Chen et al. 2021).

Arabidopsis ERF5 and ERF6, two functionally redundant TFs, are master regulators that adapt leaf growth to environmental changes. Under osmotic stress, the ethylene level is increased, which activates the expression of *MPK3* and *MPK6* in the signaling pathway. The expression of *ERF5* and *ERF6* is then induced. On the one hand, ERF5 and ERF6 TF directly activate the expression of stress-related *WRKY33*, *STZ*, and *MYB51* to trigger the stress-related signaling cascade. On the other hand, ERF5 and ERF6 transcriptionally activate the expression of the gibberellic acid (GA)-inactivating enzyme, GA2-OX6, that leads to a decrease in GA content, thereby stabilizing the content of DELLA proteins. In this way, ERF5 and ERF6 regulate leaf growth in *Arabidopsis* (Figure 5.5). *Arabidopsis* overexpressing *ERF6* exhibited dwarf phenotype and hypersensitivity to osmotic stress, while the *erf5erf6* mutants were not significantly affected by osmotic stress (Dubois et al. 2013).

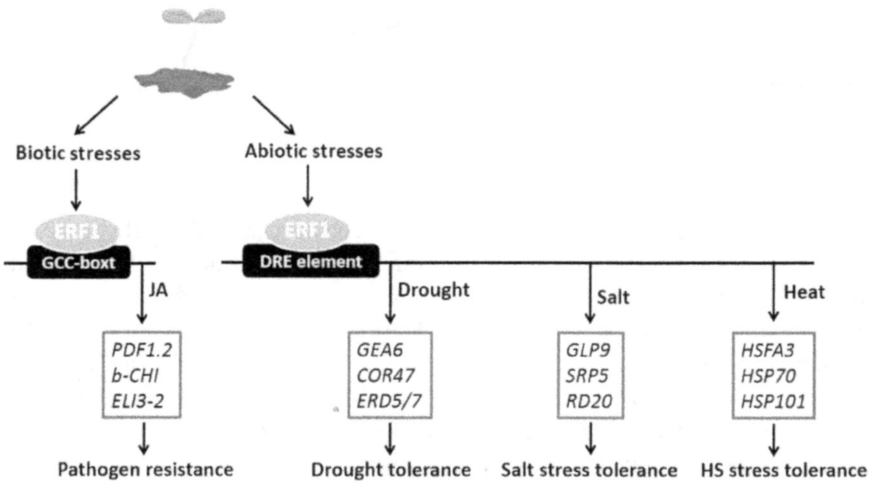

Figure 5.4 The function of EFR1 under biotic and abiotic stresses.

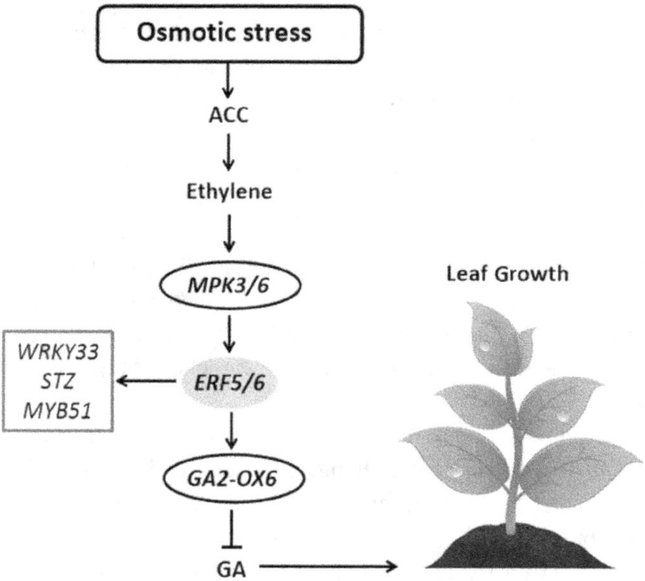

Figure 5.5 ERF5 and ERF6 TFs regulate leaf growth under osmotic stress (Deeas work/ Shutterstock).

TOMATO ETHYLENE-RESPONSIVE FACTOR2 (*TERF2*) is an allele of *LeERF2*; the proteins they encode only differ at the 253rd amino acid, with valine in *TERF2* and leucine in *LeERF2*. LeERF2/TERF2 can bind to the GCC box in the promoter of tobacco NtACS3 and the dehydration-responsive element in tomato *LeACO3*, which then transcriptionally activates the expression of ethylene biosynthesis-related genes in tobacco and tomato (Zhang et al. 2009). Under cold stress, the ethylene level is increased, which induces the expression of *TERF2/LeERF2*. TERF2/LeERF2 can transcriptionally activate *ACS/ACO* genes to further induce ethylene production and can also activate the transcription of stress-related genes to enhance freezing tolerance in plants (Figure 5.6). Overexpressing *TERF2/LeERF2* in tobacco and tomato resulted in increased freezing tolerance in transgenic plants, while antisense *TERF2/LeERF2* tomato exhibited sensitivity to freezing (Zhang and Huang 2010).

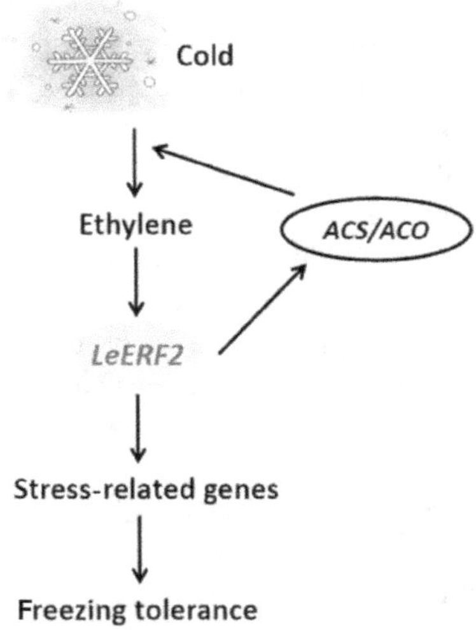

Figure 5.6 The function of TERF2/LeERF2 TF in cold stress *(adapted from Zhang et al. 2010)*.

Tomato LeERF1 was isolated by our laboratory from tomato (Zhang et al. 2005). *LeERF1* overexpression in tomato resulted in significantly increased resistance to *Rhizopus nigricans* compared to wild-type mature green and red ripe tomatoes (Figure 5.7) (Pan et al. 2013).

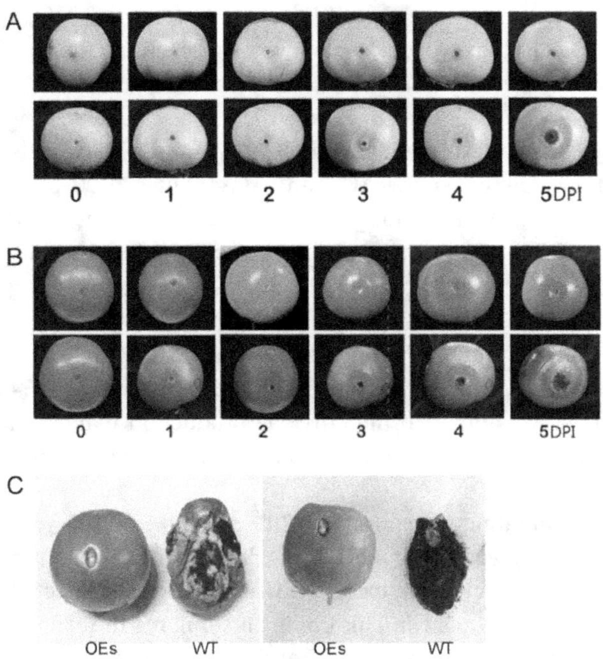

Figure 5.7 Tomato fruit overexpressing *SlERF1* exhibited increased resistance to *Rhizopus nigricans* (Pan et al 2013/with permission from ELSEVIER). Fruit of MG (A) and RR (B, C) ripening stages from *SlERF1*-OEs and WT lines were challenged by *R. nigricans*. Infection was evaluated and photographed every day from 1 to 5 DPI(A, B) or the 10th and 15th day after inoculation (C).

JASMONATE AND ETHYLENE RESPONSE FACTOR 1(JERF1) is a TF isolated from tomato using a yeast one-hybrid system with the GCC box as bait. JERF1 contains the conserved ERF domain; it can bind not only to the GCC box but also to the DEHYDRATION-RESPONSIVE ELEMENT (DRE). When treated with ethylene, JA, ABA, and salt, tomato *JERF1* expression was significantly induced, while overexpressing *JERF1* resulted in increased salt tolerance in tomato (Zhang et al. 2004a). Overexpressing *JERF1* in rice transcriptionally activated *OsABA2* and *Os03g0810800*, genes encoding two key enzymes in the ABA biosynthetic pathway, resulting in an increased ABA level and enhanced drought tolerance (Zhang et al. 2010).

5.3 The Roles of ERF Family TFs in Fruit Ripening

Fruit formation can be divided into three stages: fruit set, development, and ripening. The transition of fruit developmental stages is strictly regulated by genes and signaling pathways. Fruit ripening is a complex process regulated together by internal and external factors; it is also an area that has received considerable attention from geneticists and breeders. As fruit develops and ripens, the ethylene level increases, the fruit becomes soft, and the carotenoid level increases (Giovannoni 2004).

Accumulating studies have shown that the ERF family is critical for fruit development and ripening. During tomato fruit development and ripening, ethylene production can be divided into two distinct systems: the immature green (IMG) stage and mature green stage (MG) are in system 1; from the MG stage afterward is system 2. In system 1, ethylene synthesis mainly requires ACS1A and ACS6, while in system 2, ethylene synthesis mainly requires ACS2 and ACS4 (Liu et al. 2015).

Ethylene is crucial for tomato fruit ripening. In response to ethylene, ERF family TFs can bind to the GCC box in the promoter region and transcriptionally regulate the expression of ripening-related genes in the nucleus, thereby affecting tomato fruit ripening (Liu et al. 2015).

The expression of *ERFs* at five different ripening stages in tomato fruit was analyzed, and the results showed that the expression of *ERF.E1*, *ERF.E2*, and *ERF.E4* significantly increased after GM, similar to the ethylene production pattern in system 2, indicating that these three ERFs are involved in regulating tomato fruit ripening (Liu et al. 2015).

LeERF1-4 are four genes containing the conserved ERF domain isolated from tomato. Northern blot analysis showed that only *LeERF2*, different from the other three genes, had a high expression level in red ripe tomato fruit but was not detected in the fruit of ripening mutants *nor*, *Nr*, or *rin*. This finding is similar to the expression pattern of ripening-related gene *E8*, indicating LeERF2 is possibly involved in the regulation of tomato fruit ripening (Tournier et al. 2003). Overexpressing *LeERF1* in tomato fruit resulted in increased ethylene levels, earlier fruit softening, and a shorter shelf life. Inhibiting *LeERF1* expression in tomato led to a decreased ethylene level, increased fruit hardness, and a longer shelf life (Figure 5.8) (Li et al. 2007). Liu et al. (2018) analyzed the comprehensive expression profiling of tomato ERF. The results shown that out of the 77 ERFs present in the tomato genome, 27 show enhanced expression at the onset of ripening while 28 display a ripening associated decrease in expression, suggesting that different ERFs may have contrasting roles in fruit ripening. Silencing *SlERF6* by RNAi enhanced both carotenoid and ethylene levels during fruit ripening, demonstrating an important role for *SlERF6* in ripening (Lee et al. 2012). Overexpression of *SlERF.B3-SRDX*, a dominant repressor version of *ERF.B3*, results in a dramatic delay of the onset of ripening in tomato (Liu et al. 2014). The *SlERF.F12* is a member of ERF.F subfamily, characterized by the presence of an EAR motif. The EAR motif is the most common transcriptional repression motif identified in plants. Deng et al. (2022) demonstrated that *SlERF.F12*-mediated transcriptional repression of key ripening-related genes *ACS2, ACS4, PL,* is dependent on the presence of its C-terminal EAR motif. Furthermore, *SlERF.F12* interacts with the corepressor TPL2 via the C-terminal EAR motif and recruits histone deacetylases *SlHDA1* and *SlHDA3* to form a tripartite complex in vivo that actively represses transcription of ripening genes by decreasing the level of the permissive histone acetylation marks H3K9Ac and H3K27Ac at their promoter regions.

Studies of the effect of ethylene on apple fruit ripening have symbolic importance in plant hormone research. To understand the molecular mechanism of apple fruit ripening, scientists studied the factor that affected the rate of fruit softening in apple cultivars, namely, ethylene. Although the different ethylene production rates in these cultivars could be explained by the ACC synthase gene, the different alleles cannot explain why apple fruit exhibited different phenotypes during ripening or storage. To better understand the role of ethylene signal transduction in apple fruit ripening, scientists began to study two ERF family genes: *MdERF1* and *MdERF2*. These genes were isolated from the ripening fruit of Golden Delicious apples. Scientists chose *Malus* × *domestica* Borkh. cv. Golden

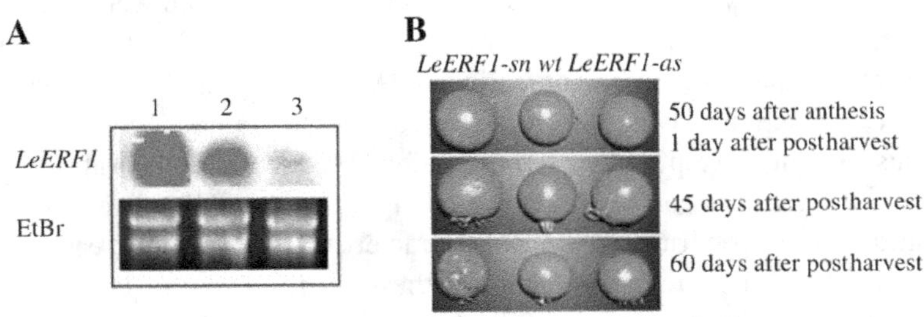

Figure 5.8 Altered shelf life of *LeERF1* transgenic tomato fruit (Li et al. 2007/with permission from Springer Nature).

Delicious as materials and studied the expression of these genes. The results showed that although *MdERF1* and *MdERF2* were both isolated from ripening apple, their expression patterns were different: *MdERF1* was predominantly expressed in ripening fruit, while *MdERF2* was exclusively expressed in ripening fruit. Moreover, the abundance of *MdERF1* was significantly higher than that of *MdERF2*. When treated with ripening inhibitor 1-MCP, the expression of *MdERF1* and *MdERF2* in ripening fruit was inhibited. In the Fuji cultivar that had low ethylene production, *MdERF1* and *MdERF2* had low expression levels in ripening fruit. In contrast, in Golden Delicious, which had high ethylene production, *MdERF1* and *MdERF2* expression levels were high, and their expression patterns were consistent with those of other ripening-related genes, *MdACS1*, *MdACO1*, and *MdPG1*, in these two cultivars. These results indicate that transcription factors MdERF1 and MdERF2 are associated with apple fruit ripening (Wang et al. 2007).

During papaya fruit ripening, the expression of *CpERF9* decreased. Propylene treatment resulted in earlier ripening, and the decrease in *CpERF9* expression also occurred earlier. The expression of the pectin methylesterase gene *CpPME1/2* and polygalacturonase gene *CpPG5* increased, while papaya fruit firmness declined. Electrophoresis mobility shift assay (EMSA) analysis showed that CpERF9 could bind to the GCC box of the promoter of *CpPME1*, *CpPME2*, and *CpPG5* in vitro, thereby regulating papaya fruit firmness and affecting papaya fruit ripening (Fu et al. 2016).

Fourteen *AdERF* genes have been cloned in kiwifruit, most of which are found to be expressed during early stages of fruit ripening, such as *AdERF4*, *AdERF6*, *AdERF7*, *AdERF8*, *AdERF10*, *AdERF12*, and *AdERF14*. Some *AdERFs* are clearly under positive control of ethylene, such as *AdERF4*, *AdERF6*, *AdERF10*, and *AdERF14*. Ethylene treatment led to increased expression of these genes, while ethylene inhibitor 1-MCP treatment resulted in reduced expression, indicating that they are associated with fruit ripening in kiwifruit (Yin et al. 2010).

Seven *Ps-ERFs* were cloned from two plum cultivars—Early Golden (EG) and Shiro (SH)—and their expression patterns during fruit development were studied. 1-MCP treatment showed that *Ps-ERF1a*, *Ps-ERF2b*, *Ps-ERF3a*, *Ps-ERF3b*, and *Ps-ERF12* were positively regulated by ethylene in both cultivars and were likely involved in plum fruit development and ripening (El-Sharkawy et al. 2009).

Banana is one of the most important fruit varieties in tropical and subtropical countries; the production of banana ranks the second highest among world fruit crops. It is also among the top 10 most important food products in the world. As a result, banana has received considerable research attention. In addition, as a typical respiration climacteric fruit, banana has characteristics of a rapid increase in respiration intensity and increased ethylene synthesis during ripening. From the economic perspective, this characteristic also leads to a short shelf life. Although many external methods have been used to extend the shelf life of banana, including postharvest refrigeration, heat treatment, and chemical and biological treatment, they cannot ultimately change the reality of severe postharvest damage in banana. Thus, it is necessary to first understand the regulators of ripening in banana in order to develop a more effective method for postharvest storage. Based on the fact that banana is a respiration climacteric fruit, tremendous effort has been made in studying ethylene biosynthesis and substances related to ethylene signal transduction in banana, including ACC synthase (ACS), ACC oxidase (ACO), and ethylene receptors. Subsequently, scientists have cloned 15 *MaERF* genes from banana fruit. Through treatments using 1-MCP, ethylene, and 1-MCP combined with ethylene, it has been found that *MaERF9* is positively regulated by ethylene, while *MaERF11* is negatively regulated by ethylene. Yeast one-hybrid and transient expression analyses showed that MaERF11 could bind to the promoter of *MaACS1* and *MaACO1* and inhibit their expression, while MaERF9 could only bind to the promoter of *MaACS1* and induce its expression (Xiao et al. 2013). These results indicate that MaERFs are likely involved in banana fruit ripening.

Banana MaDEAR1 belongs to the DREB subfamily of ERF TFs and is a transcriptional repressor negatively regulated by ethylene. EMSA analysis showed that MaDEAR1 could directly bind to cell wall-modifying genes, such as *MaEXP1/3*, *MaPG1*, *MaXTH10*, *MaPL3*, and *MaPME3* and inhibit their expression, thereby affecting fruit softening in banana (Fan et al. 2016). Kuang et al. (2017) identified four genes encoding ripening-regulated DREB TFs in banana, *MaDREB1*, *MaDREB2*, *MaDREB3*, and *MaDREB4*, and demonstrated that they play regulatory roles in fruit ripening. Among the four *DREB* TFs, *MaDREB2* is induced and upregulated by exogenous or endogenous ethylene during fruit ripening, and the activation of MaDREB2 may initiate a transcriptional regulatory cascade or form a positive feedback loop to auto-amplify its own transcription, thus directly or indirectly modulate the ripening of banana fruit.

Grape, also one of the most important crops in the world, is an important resource. It can be consumed directly as a fresh fruit or used in fermentation as the raw material for wine or other beverages. During the ripening process in grape, in order to maintain the color, sweetness, and flavor of the berries, many transcriptional and metabolic changes occur. At the beginning of ripening, the metabolism of grape changes dramatically, which further impacts sugar accumulation, acid metabolism, and the flavor and color of the berries. Unlike banana, grape is not a respiration climacteric fruit. Grape *VviERF045* is only expressed before ripening and reaches its peak at ripening. Grape leaves overexpressing *VviERF045* had lower levels of chlorophyll and carotenoids as well as decreased photosynthetic capability, indicating that *VviERF045* may be involved in regulating grape ripening (Leida et al. 2016).

A comparison of the transcriptomes between a late-ripening sweet orange mutant (Fengwan, MT) and the wild type (Fengjie 72–1) revealed that the number of *ERFs* was the greatest among differentially expressed TFs, indicating that ERFs are involved in the ripening of sweet orange (Wu et al. 2016). In citrus, *CitERF13* is a transcriptional activator, associated with fruit degreening, via transactivation on chlorophyll degradation-related genes, such as *CitPPH* (Yin et al. 2016).

In summary, ERF TFs are important regulators and play crucial roles in plant growth, development, and fruit ripening.

Ethylene response factor (AP2/ERF) is an important regulator in plants, with 146 AP2/ERF transcription factors in tomatoes. ERF transcription factors are involved in the entire life activity process, such as development, maturation, biological and abiotic stress. When infected with pathogenic bacteria, ERF transcription factors bind to the GCC box of the gene, while in the case of abiotic stress, ERF1 binds to DRE elements (G/ACCGAC). However, their function as a regulatory mechanism in plants subjected to biological/abiotic stress remains poorly understood.

References

An, J.P., Zhang, X.W., Bi, S.Q. et al. (2020). The ERF transcription factor MdERF38 promotes drought stress-induced anthocyanin biosynthesis in apple[J]. *Plant J* 101 (3): 573–589.

Chakravarthy, S., Tuori, R.P., D'Ascenzo, M.D. et al. (2003). The tomato transcription factor Pti4 regulates defense-related gene expression via GCC box and non-GCC box cis elements. *Plant Cell* 15: 3033–3050.

Charfeddine, M., Saidi, M.N., Charfeddine, S. et al. (2015). Genome-wide analysis and expression profiling of the ERF transcription factor family in potato (Solanum tuberosum L.). *Mol Biotechnol* 57: 348–358.

Chen, Y., Zhang, L., Zhang, H. et al. (2021). ERF1 delays flowering through direct inhibition of FLOWERING LOCUS T expression in Arabidopsis. *J Integr Plant Biol* 63: 1712–1723.

Cheng, M.C., Liao, P.M., Kuo, W.W., and Lin, T.P. (2013). The Arabidopsis ETHYLENE RESPONSE FACTOR1 regulates abiotic stress-responsive gene expression by binding to different cis-acting elements in response to different stress signals. *Plant Physiol* 162: 1566–1582.

Deng, H., Chen, Y., Liu, Z. et al. (2022). SlERF. F12 Modulates the Transition to Ripening in Tomato Fruit by Recruiting the Co-repressor TOPLESS and Histone Deacetylases to Repress Key Ripening genes[J]. *Plant Cell* 34 (4): 1250–1272.

Dubois, M., Skirycz, A., Claeys, H. et al. (2013). Ethylene response Factor6 acts as a central regulator of leaf growth under water-limiting conditions in Arabidopsis. *Plant Physiol* 162: 319–332.

El-Sharkawy, I., Sherif, S., Mila, I. et al. (2009). Molecular characterization of seven genes encoding ethylene-responsive transcriptional factors during plum fruit development and ripening. *J Exp Bot* 60: 907–922.

Fan, Z.Q., Kuang, J.F., Fu, C.C. et al. (2016). The banana transcriptional repressor MaDEAR1 negatively regulates cell wall-modifying genes involved in fruit ripening. *Front Plant Sci* 7: 1021.

Fu, C.C., Han, Y.C., Qi, X.Y. et al. (2016). Papaya CpERF9 acts as a transcriptional repressor of cell-wall-modifying genes CpPME1/2 and CpPG5 involved in fruit ripening. *Plant Cell Rep* 35: 2341–2352.

Giovannoni, J.J. (2004) Genetic regulation of fruit development and ripening. 16: S170–S180.

Hu, Y., Chang, C., Xu, G., and Wang, T. (2011). Light restored root growth of Arabidopsis with constitutive ethylene response. *Acta Physiologiae Plantarum* 33: 667–674.

Hu, Y., Zhao, L., Chong, K., and Wang, T. (2008) Overexpression of OsERF1, a novel rice ERF gene, up-regulates ethylene-responsive genes expression besides affects growth and development in Arabidopsis. J Plant Physiol 165: 1717–1725.

Ito, T.M., Polido, P.B., Rampim, M.C. et al. (2014). Genome-wide identification and phylogenetic analysis of the AP2/ERF gene superfamily in sweet orange (Citrus sinensis). *Genet Mol Res* 13: 7839–7851.

Jofuku, K.D., Montagu, M.V., and Okamuro, J.K. (1994). Control of Arabidopsis flower and seed development by the homeotic gene APETALA2. *Plant Cell* 6: 1211–1225.

Ju, C. and Chang, C. (2015). Mechanistic insights in ethylene perception and signal transduction. *Plant Physiol* 169: 85–95.

Komatsu, M., Chujo, A., Nagato, Y. et al. (2003). FRIZZY PANICLE is required to prevent the formation of axillary meristems and to establish floral meristem identity in rice spikelets. *Development* 130: 3841–3850.

Kuang, J.F., Chen, J.Y., Liu, X.C. et al. (2017). The transcriptional regulatory network mediated by banana (Musa acuminata) dehydration-responsive element binding (MaDREB) transcription factors in fruit ripening[J]. *New Phytol* 214 (2): 762–781.

Lee, J.M., Joung, J.G., McQuinn, R. et al. (2012). Combined transcriptome, genetic diversity and metabolite profiling in tomato fruit reveals that the ethylene response factor SlERF6 plays an important role in ripening and carotenoid accumulation[J]. *Plant J* 70 (2): 191–204.

Leida, C., Dal Ri, A., Dalla, C.L. et al. (2016). Insights into the role of the berry-specific ethylene responsive factor VviERF045. *Front Plant Sci* 7: 1793.

Li, Y., Zhu, B., Xu, W. et al. (2007). LeERF1 positively modulated ethylene triple response on etiolated seedling, plant development and fruit ripening and softening in tomato. *Plant Cell Rep* 26.

Licausi, F., Giorgi, F.M., Zenoni, S. et al. (2010). Genomic and transcriptomic analysis of the AP2/ERF superfamily in Vitis vinifera. *BMC Genomics* 11: 719.

Liu, M., Chen, Y., Chen Y, et al. (2018). The tomato Ethylene Response Factor Sl-ERF. B3 integrates ethylene and auxin signaling via direct regulation of Sl-Aux/IAA27. *New Phytol* 219: 631–640.

Liu, M., Diretto, G., Pirrello, J. et al. (2014). The chimeric repressor version of an E thylene response factor (ERF) family member, Sl-ERF. B3. *Shows Contrasting Effects on Tomato Fruit ripening[J]. New Phytologist* 203 (1): 206–218.

Liu, M., Pirrello, J., Chervin, C. et al. (2015). Ethylene control of fruit ripening: revisiting the complex network of transcriptional regulation. *Plant Physiol* 169: 2380–2390.

Mehrnia, M., Balazadeh, S., Zanor, M.I., et al. (2013). EBE, an AP2/ERF transcription factor highly expressed in proliferating cells, affects shoot architecture in Arabidopsis. *Plant Physiol* 162: 842–857.

Nakano, T., Suzuki, K., Fujimura, T., and Shinshi, H. (2006). Genome-wide analysis of the ERF gene family in Arabidopsis and rice. *Plant Physiol* 140: 411–432.

Ohme-Takagi, M. and Shinshi, H. (1995). Ethylene-inducible DNA binding proteins that interact with an ethylene-responsive element. *Plant Cell* 7: 173–182.

Pan, X.Q., Fu, D.Q., Zhu, B.Z. et al. (2013). Overexpression of the ethylene response factor SlERF1 gene enhances resistance of tomato fruit to Rhizopus nigricans. *Postharvest Biol Tec* 75: 28–36.

Pandey, G.K., Grant, J.J., Cheong, Y.H. et al. (2005). ABR1, an APETALA2-domain transcription factor that functions as a repressor of ABA response in Arabidopsis. *Plant Physiol* 139: 1185–1193.

Pirrello, J., Jaimes-Miranda, F., Sanchez-Ballesta, M.T. et al. (2006). Sl-ERF2, a tomato ethylene response factor involved in ethylene response and seed germination. *Plant Cell Physiol* 47: 1195–1205.

Pirrello, J., Prasad, B.C.N., Zhang, W. et al. (2012). Functional analysis and binding affinity of tomato ethylene response factors provide insight on the molecular bases of plant differential responses to ethylene. *BMC Plant Biol* 12: 190.

Pre, M., Atallah, M., Champion, A. et al. (2008). The AP2/ERF domain transcription factor ORA59 integrates jasmonic acid and ethylene signals in plant defense. *Plant Physiol* 147: 1347–1357.

Qu, X., B P, H., Gao, Z. et al. (2007). A strong constitutive ethylene-response phenotype conferred on Arabidopsis plants containing null mutations in the ethylene receptors ETR1 and ERS1. *BMC Plant Biol* 7 (1): 1–15.

Sakuma, Y., Liu, Q., Dubouzet, J.G. et al. (2002). DNA-binding specificity of the ERF/ AP2 domain of Arabidopsis DREBs, transcription factors involved in dehydration- and cold-inducible gene expression. *Biochem Biophys Res Commun* 290: 998–1009.

Sasaki, K., Mitsuhara, I., Seo, S. et al. (2007). Two novel AP2/ERF domain proteins interact with cis-element VWRE for wound-induced expression of the Tobacco tpoxN1 gene. *Plant J* 50: 1079–1092.

Tournier, B., Sanchez-Ballesta, M.T., Jones, B. et al. (2003). New members of the tomato ERF family show specific expression pattern and diverse DNA-binding capacity to the GCC box element. *FEBS Lett* 550: 149–154.

van der Graaff, E., Dulk-Ras, A.D., Hooykaas, P.J., and Keller, B. (2000). Activation tagging of the LEAFY PETIOLE gene affects leaf petiole development in Arabidopsis thaliana. *Development* 127: 4971–4980.

Wang, A., Tan, D., Takahashi, A. et al. (2007). MdERFs, two ethylene-response factors involved in apple fruit ripening. *J Exp Bot* 58: 3743–3748.

Wu, J., Fu, L., and Yi, H. (2016). Genome-wide identification of the transcription factors involved in citrus fruit ripening from the transcriptomes of a late-ripening sweet orange mutant and its wild type. *PLoS One* 11: e154330.

Xiao, Y.Y., Chen, J.Y., Kuang, J.F. et al. (2013). Banana ethylene response factors are involved in fruit ripening through their interactions with ethylene biosynthesis genes. *J Exp Bot* 64: 2499–2510.

Yin, X., Xie, X., Xia, X. et al. (2016). Involvement of an ethylene response factor in chlorophyll degradation during citrus fruit degreening[J]. *Plant J* 86 (5): 403–412.

Yin, X.R., Allan, A.C., Chen, K.S., and Ferguson, I.B. (2010). Kiwifruit EIL and ERF genes involved in regulating fruit ripening. *Plant Physiol* 153: 1280–1292.

Zhang, H., Huang, Z., Xie, B. et al. (2004a). The ethylene-, jasmonate-, abscisic acid- and NaCl-responsive tomato transcription factor JERF1 modulates expression of GCC box-containing genes and salt tolerance in tobacco. *Planta* 220: 262–270.

Zhang, H., Yang, Y., Zhang, Z. et al. (2008). Expression of the ethylene response factor gene TSRF1 enhances abscisic acid responses during seedling development in tobacco. *Planta* 228: 777–787.

Zhang, H., Zhang, D., Chen, J. et al. (2004b). Tomato stress-responsive factor TSRF1 interacts with ethylene responsive element GCC box and regulates pathogen resistance to Ralstonia solanacearum. *Plant Mol Biol* 55: 825–834.

Zhang, H., Zhu, B., Yu, B. et al. (2005). Cloning and DNA-binding properties of ethylene response factor, LeERF1 and LeERF2, in tomato. *Biotechnol Lett* 27: 423–428.

Zhang, L., Liu, P., Wu, J. et al. (2020). Identification of a novel ERF gene, TaERF8, associated with plant height and yield in wheat[J]. *BMC Plant Biol* 20 (1): 1–12.

Zhang, Z. and Huang, R. (2010). Enhanced tolerance to freezing in tobacco and tomato overexpressing transcription factor TERF2/LeERF2 is modulated by ethylene biosynthesis. *Plant Mol Biol* 73: 241–249.

Zhang, Z., Li, F., Li, D. et al. (2010). Expression of ethylene response factor JERF1 in rice improves tolerance to drought. *Planta* 232: 765–774.

Zhang, Z., Zhang, H., Quan, R. et al. (2009). Transcriptional regulation of the ethylene response factor LeERF2 in the expression of ethylene biosynthesis genes controls ethylene production in Tomato and Tobacco. *Plant Physiol* 150: 365–377.

Zhuang, J., Peng, R.H., Cheng, Z.M.M. et al. (2009). Genome-wide analysis of the putative AP2/ERF family genes in Vitis vinifera. *Sci Hortic* 123: 73–81.

6

The ARF Side of the Fruit Tuning of Flesh Fruit Development and Ripening

6.1 Introduction

Auxin is an important hormone family involved in the growth and development of plants. Studies in recent years have revealed the important mechanisms of auxin perception, signal transduction, and transport, making it the most intensively studied hormone family in plants (Benjamins and Scheres 2008). With the deepening of research, scientists have identified genes and gene families involved in auxin signal transduction. Among them, two transcription factor families, the auxin/indole-3-acetic acid (Aux/IAA) transcription factor family and the auxin response factor (ARF) family, are crucial in auxin signal transduction (Woodward and Bartel 2005; Hagen 2015). This chapter is focused on the role of the ARF family in regulating fruit development and ripening.

6.2 Overview of the ARF Family

6.2.1 Discovery of ARF

ARF was first discovered by American scientists. Through screening the *Arabidopsis* cDNA library by using a yeast one-hybrid system with artificially synthesized AuxREP3 (4X) as a molecular bait, researchers isolated the *ARF* gene for the first time and named it *ARF1*. This discovery was published in *Science* in 1997 (Ulmasov et al. 1997) and laid a foundation for the subsequent discoveries of *ARF*s in other species. During the past two decades, *ARF* genes in rice (*Oryza sativa*) (Sato et al. 2001; Waller et al. 2002), wheat (Kobayashi-Uehara et al. 2001), maize (*Zea mays* L.) (Xing et al. 2011), tomato (*Solanum lycopersicum*) (Kumar et al. 2011), soybean (Le et al. 2016), *Populus trichocarpa* (Kalluri et al. 2007), banana (*Musa acuminata* L.) (Hu et al. 2015), and sweet orange (*Citrus sinensis*) (Li et al. 2015) have been identified.

Transcriptional Regulation of Flesh Fruit Development and Ripening, First Edition. Yunbo Luo.

The chromosomal distribution of *ARF* genes is described below: The *Arabidopsis* genome contains 23 *ARF* genes on five chromosomes, among which chromosome I has 13 *ARF* genes, the largest number, and chromosomes II–IV have three, one, two, and four *ARF* genes, respectively (Figure 6.1). As the chromosomal distribution shows, *ARF12, 13, 14, 15, 20, 21*, and *22* can form a cluster of *ARF* genes on chromosome I. These genes share a high degree of similarity among their amino acid and corresponding protein-coding nucleotide sequences, suggesting that these genes may originate from the same tandem copy. Phylogenetic analysis of the 23 ARF family members revealed several closely related pairs of *ARF*s, namely, ARF1/ARF2, ARF3/ARF4, ARF6/ARF8, ARF7/ARF19, ARF11/ARF18, and ARF10/ARF16. This pattern suggests that these ARF members might function redundantly (Okushima et al. 2005).

The tomato *ARF* gene family contains 21 members, namely, *ARF1, 2, 3, 4, 5, 6, 6–1, 7, 8, 8–1, 9, 10, 11, 12, 13, 13–1, 14, 16, 17, 19*, and *19–1*. These genes are located on all 11 chromosomes except chromosome 9. Chromosomes 5, 7, and 11 each have three *AFRs*; chromosomes 2, 3, 8, and 12 each have two *AFRs*; and chromosomes 1, 4, 6, and 10 each have one *AFR* member (Figure 6.2). It is worth noting that although the members in each pair of *ARF6/ARF6-1*, *ARG8/ARF8-1*, and *ARF19/ARF19-1* have a relatively high homology, they are not located on the same chromosome, which may be attributed to gene duplication events (Wu et al. 2011).

ARF genes are represented by a large multigene family in plants. Bioinformatics analysis shows that ARF members have certain homology among different species. For example, 48 *ARF* genes in *Arabidopsis thaliana* (At) and *Oryza sativa* (Os) can be divided into three classes, including 13 sister pairs, namely, 9 *OsARF-OsARF*, 2 *AtARF-AtARF*, and 2 *AtARF-OsARF* sister pairs. The 79 *ARF* genes in *Zea mays* (Zm), *Arabidopsis thaliana* (At), and *Oryza sativa* (Os) can be divided into 6 classes,

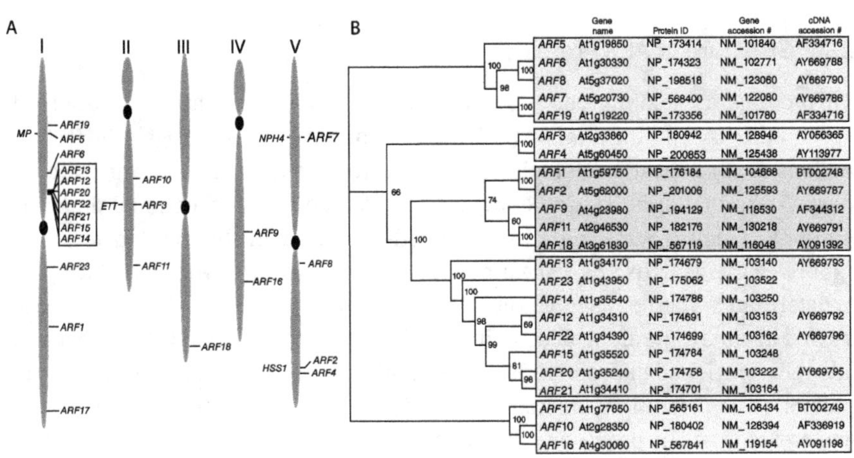

Figure 6.1 Chromosomal distribution and phylogenetic relationship of *ARF* genes in *Arabidopsis*.

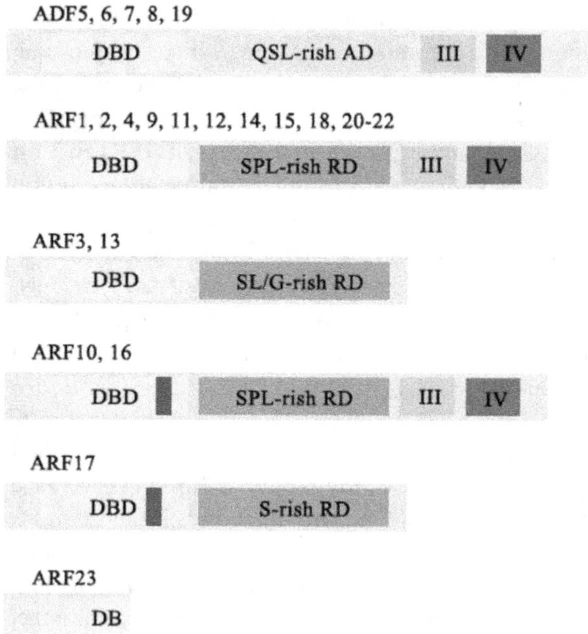

Figure 6.2 Structural domains of ARF proteins in *Arabidopsis*.

including 9 *ZmARF-ZmARF*, 11 *ZmARF-OsARF*, and 4 *AtARF-AtARF* sister pairs. The 40 *ARF* genes in *Solanum lycopersicum* (Sl) and *Arabidopsis thaliana* (At) can be divided into three major classes, including seven SlARF-AtARF, three SlARF-SlARF, and four AtARF-AtARF sister pairs (Kumar et al. 2011).

6.2.2 Structure of ARF Proteins

ARF proteins are mainly composed of three functional domains: a conserved N-terminal DNA-binding domain (DBD), a nonconserved middle region (MR) that functions as an activation domain (AD) or repression domain (RD), and a conserved C-terminal dimerization domain (CTD, including the functional domains III and IV) (Figure 6.2). The DBD is directly involved in the binding to auxin response elements (AuxREs). The amino acid sequence of the MR in an ARF protein determines whether its function is in activation or repression. Generally, those enriched in glutamine along with serine and leucine residues are mostly activators, while those enriched in proline/glycine, serine, and threonine residues are mostly repressors. The C-terminal functional domain enhances the control of ARF over its downstream genes by promoting dimerization (Le et al. 2016).

The DBD of the ARF protein belongs to a plant-specific B3-type DBD. This domain requires an addition of a large number of amino and carboxy-terminal amino acids for efficient in vitro binding to TGTCTC/GAGACA AuxREs (Tiwari et al. 2003). As shown by nuclear magnetic resonance (NMR) tests on the molecular structure of ARF, the B3-type DBD region consists of a seven-stranded open β-barrel and two α-helices located at the ends of the barrel, and this structure is extremely similar to the structure of the DBD of the restriction endonuclease EcoRII in *Escherichia coli* (Yamasaki et al. 2004). In *Arabidopsis thaliana*, all 23 AtARF proteins have a DBD. However, it is worth noting that AtARF23 contains only a DBD and is structurally incomplete as it does not have an MR or CTD. Therefore, AtARF23 is considered a pseudogene (Guilfoyle and Hagen 2007). In *Oryza sativa*, OsARF20 has two DBDs, and all other 24 ARFs contain a typical DBD (Wang et al. 2007). *Zea mays* has 31 ZmARF proteins, all of which contain a DBD at the N-terminus (Xing et al. 2011).

In an ARF protein, the MR determines whether its transcriptional function is in activation or repression. The MR can be divided into an AD and RD based on its amino acid sequence. Among the 23 ARF proteins in *Arabidopsis thaliana*, ARF5–8 and ARF19, whose MRs are rich in glutamine, serine, and leucine, are transcriptional activators. Other ARF members, with the exception of ARF23, may be transcriptional repressors that inhibit the transcription of auxin response genes. This function has been experimentally tested in transfected protoplasts. However, the tests relied on transient expression assays of ARF members in leaf mesophyll or suspension cell culture protoplasts, and therefore, the experimental design was not very rigorous. As a result, it remains possible that an ARF classified as a repressor could function as an activator or an ARF classified as an activator could function as a repressor in certain cell types or environments (Guilfoyle and Hagen 2007). In other words, ARF might have dual functions of activation and repression.

The structure and function of the CTD of ARF proteins are similar to those of domains III and IV in Aux/IAA proteins. All *Arabidopsis* ARFs with the exceptions of ARF3, 13, and 17 have a conserved CTD. Without a CTD, ARF3, 13, and 17 can still be transfected into protoplasts, indicating that the binding of an ARF activator to AuxREs does not require auxin, whereas an auxin response requires the ARF CTD (Tiwari et al. 2003). In *Oryza sativa*, all ARFs except ARF2, 3, 13, 14, 15, and 20 contain a CTD (Wang et al. 2007). All ZmARFs other than ZmARF5 and 31 contain a CTD at the C-terminus. Bimolecular fluorescence complementation and yeast two-hybrid experiments have confirmed that the CTD is involved in regulating ARF–ARF, ARF–Aux/IAA, and Aux/IAA–Aux/IAA interactions (Korasick et al. 2015).

6.2.3 Gene Structure and Phylogenetic Analysis of *ARFs*

Wang et al. (2007) analyzed the structure and phylogenetic relationships of *ARF* genes in *Oryza sativa*. They first aligned the full-length cDNA sequence with the corresponding genomic DNA sequence to obtain the number and location of exons and introns of each *ARF* gene. The results showed that all *ARF* members contain 2–14 introns. The unrooted phylogenetic tree generated from the alignments of the amino acid sequences of all the *ARFs* showed that all the *OsARF* genes could be divided into three classes: Class I, II, and III. Class I contained six members: *ARF1*, *4*, *7*, *9*, *23*, and *24*. Class II was further divided into two subclasses—IIa and IIb—containing nine and four *ARFs*, respectively. Class III contained six *ARFs*. To further elucidate the relationship between *AtARFs* and *OsARFs*, a phylogenetic tree with 25 OsARFs and 23 AtARFs was constructed. These 48 *ARF* genes were also divided into three classes—Class I, II, and III—containing 19, 20, and 9 *ARF* genes, respectively. Class I and Class II were each further divided into two subclasses (Wang et al. 2007).

Using bioinformatics methods, Zheng et al. (2021) identified 46 *ARF* genes in sorghum [*Sorghum bicolor* (L.) Moench]. According to their genetic relationship with *AtARFs*, the *SbARF* genes were divided into four classes (I–IV). Each of the *SbARF* genes consists of 1–13 introns located in three highly conserved regions of the ARF, B3, and Aux/IAA domains (Zheng et al. 2021). A total of 36 *ARF* genes were identified and validated from the maize genome through an iterative strategy. These genes are distributed on all chromosomes except chromosome 7, and all have one B3 DBD consisting of seven-stranded β sheets and two short α helixes. Phylogenetic analysis of the above genes revealed that 36 ARF proteins could be divided into three classes, among which Class I consisted of 21 *ARFs* (Wang et al. 2012). The *ARF* genes in cucumber have also been identified, and their gene structure and phylogenetic relationships have been analyzed. Fifteen *ARFs* distributed on seven chromosomes were identified using the AtARF protein sequence to search the cucumber genome database. Further, phylogenetic tree analysis was performed based on the simultaneous alignment of 15 cucumber ARF protein sequences with 23 AtARF, 31 ZmARF, and 25 OsARF protein sequences. The results showed that the 15 cucumber ARF proteins did not have high similarity, but some *ARF* genes showed a relatively high similarity among species (Liu and Hu 2013).

Luo et al. (2014) used the amino acid sequences encoded by the *Arabidopsis ARF* gene family as queries for a BLAST search against the Apple GFDB database (Apple Gene Function and Gene Family Database) and obtained the candidate *ARF* genes in apple (*Malus domestica*). By using conserved domain prediction programs Pfam and SMART to verify the presence of ARF structural domains, 31 apple *ARF* genes were finally identified. The intron and exon distribution maps of *ARF* genes were obtained based on the apple genome data downloaded from the Apple GFDB database. The analysis revealed that the apple ARF family members were composed of multiple exons, and the open reading frame length ranged from 816 to 3450 bp. Meanwhile, a phylogenetic analysis of 31 ARF protein sequences by using ClustalX software revealed that apple ARF proteins could be divided into four groups in terms of homology, and 13 sister pairs of paralogous *ARF* genes were found (Luo et al. 2014). Using the method described above, Tombuloglu analyzed the *ARF* genes in barley (*Hordeum vulgare* L.) and finally identified 20 genes that were distributed on six chromosomes but not on chromosome 4. According to the phylogenetic tree constructed on 20 barley ARFs, 36 ZmARFs, 25 OsARFs, 31 rape ARFs, 23 AtARFs, 21 SlARFs, and 19 grape ARFs, the ARF proteins in these seven plants could be divided into six groups (Tombuloglu 2019).

Forty-nine ARFs were obtained by HMM search and BLAST analysis in banana genome database. MaARF1 to MaARF49 were distributed on all chromosomes. Based on transcriptome sequencing data, Kong et al. further screened out those members with no or very low expression, resulting in 35 ARFs. These members were divided into four groups, and all ARFs had typical ARF protein structure, which is composed of a highly conserved DBD in the N-terminal region with a plant-specific B3-type subdomain and an auxin-resp subdomain. In addition, 25 ARFs had the AUX_IAA dimerization subdomains, indicating that these ARFs can interact with each other or Aux/IAA proteins. The full length of these banana ARF proteins ranged from 584 to 1137 aa, and their relative molecular mass varied from 63.15 to 126.75 kDa, with PIs from 5.39 to 8.54 (Kong et al. 2022).

A total of 21 ARF genes were confirmed by a BLASTp search against the cannabis genome and detection of the DBD with the NCBI-CDD tools. All 21 CsARF genes were positioned on cannabis chromosomes. The ARF proteins ranged from 402 amino acids (aa) (CsARF16) to 1148 aa (CsARF5) in length; MV ranged from 45.40 (CsARF16) to 128.29 kDa. Phylogenetic relationships showed that 21 ARF proteins were divided into 3 classes (class I, class II, and class III), and the members of the same subfamily had similar gene structures . The majority of CsARFs contained more than 10 exons, and the class III family members contained only 2–5 exons. In addition, 15 variable splicing events occurred in 7 ARF genes (Zhang et al. 2022).

Based on the CDS sequences of the ARF genes from apple, grape, tomato, Arabidopsis, and rice, 60 VcARF genes were totally identified. In brief, the CDS lengths of VcARF family vary from 1698 to 6417, and their deduced proteins comprise 565–2138 amino acids with the estimated molecular weights from 62.77 to 236.27 kDa. Among them, 9 digenic, 5 trigenic, and 6 tetragenic VcARF pairs exhibited more than 95% identity to each other. Phylogenetic analysis was performed using the CDS sequences of VcARFs and the ARFs from grape, apple, tomato, Arabidopsis, and rice. Consequently, all the ARFs were classified into five different groups (Li et al. 2022).

By querying tomato ARFs against the eggplant genome newly reported using BLASTP, a total of 20 candidate genes were identified. Of the identified genes, SmARF19 and SmARF16A proteins contain the longest (1423 aa) and shortest (379 aa) amino acid sequences. The molecular weight of the SmARF proteins ranged from 43.09 to 158.50 kDa. All SmARF proteins were found to contain an ARF domain. The phylogenetic tree was constructed using ARF protein sequences from eggplant, Arabidopsis, and tomato. In total, the 67 sequences included in the phylogenetic tree were grouped into three major clades (Chen et al. 2022).

6.2.4 Expression Patterns of *ARF* Genes

In recent years, studies have investigated the expression characteristics of *ARF* genes and found that the ARF family consists of a larger number of members with various spatiotemporal expression patterns in plants.

In *Arabidopsis thaliana*, most *ARF* genes are specifically expressed in roots, stems, leaves, flowers, fruits, and seeds. *ARF1* expression is high in flowers but low or even absent in other parts. *ARF2* is expressed in all tissues (Ellis et al. 2005; Iwasaki et al. 2013). *ARF3* is expressed at a low level in apical meristems and a high level in leaf primordia (Iwasaki et al. 2013). *ARF4* is expressed in the abaxial region of leaf primordia, especially in the leaf base (Pekker et al. 2005). *ARF6* and *ARF8* are expressed in certain parts of flowers, including the filament, anther, and pistil (Wu et al. 2006; Goetz et al. 2007). *ARF7* and *ARF19* are expressed in seedlings, lateral roots, and developing embryos (Wang et al. 2005b; Okushima et al. 2007). *ARF10* and *ARF16* are expressed in embryos, root caps, root meristems, and leaves (Wang et al. 2005a).

In *Solanum lycopersicum*, *ARF3*, *ARF5*, *ARF6*, *ARF9*, *ARF10*, *ARF13*, *ARF15*, and *ARF17* have similar functions, and their expression levels are the highest in the fruits at the development stage. *ARF1*, *ARF9*, *ARF11*, *ARF15*, and *ARF16* are significantly expressed in the flowers at the development stage; among them, *ARF16* is the most highly expressed (Kumar et al. 2011). *ARF12* is mainly expressed in leaves and hypocotyl but not roots or reproductive organs (Jones et al. 2002).

In *Oryza sativa*, 24 of the 25 *OsARF* genes are expressed in roots, stems, leaves, and young panicles, while *OsARF20* is not expressed in seedlings. The expression levels of most *OsARFs* do not fluctuate significantly within tissues (Wang et al. 2007).

In potato tubers (*Solanum tuberosum*), *ARF6* is highly expressed in the peripheral zone of the apical meristem in the early stages of sprouting but undetectable in dormant buds. *ARF6* expression strongly increased when apical dominance was removed (Faivre-Rampant et al. 2004).

Wen et al. (2019) analyzed the expression characteristics of 67 *ARF* genes in different tissues and organs in *Brassica napus* based on RNA-Seq data. They found that *ARFs* have a broad expression profile, with the highest expression level in roots (Wen et al. 2019). *ARF* gene expression in pepper (*Capsicum annuum*) has also been studied. The results showed that the expression levels of *ARF1*, *ARF2B*, *ARF5*, *ARF6*, *ARF9B*, *ARF16B*, and *ARF19A* were the highest in roots; the highest transcription levels of *ARF3*, *ARF4*, *ARF7*, *ARF8A*, and *ARF8B* were in the hypocotyl; and the highest expression levels of *ARF2A* and *ARF9A* were in buds (Zhang et al. 2017). Tang et al. (2018) examined the expression levels of *ARF* genes in different tissues in physic nut (*Jatropha curcas* L.). They found that, among 17 *ARF* genes, *ARF16* had the highest expression level in roots and *ARF10* had the highest expression level in seeds (Tang et al. 2018). The fragments per kilobase million (FPKM) values of 21 ARF genes in the flowers, bracts, leaves, stems, roots, and seeds in cannabis were obtained by transcriptomic analysis. The data showed that the gene expression levels of the CsARF genes involved in the growth and development of cannabis were significantly different in various tissues and organs. Twelve genes were expressed at a high level in the flower and bract tissues and five genes were expressed at a high level in all tissues. CsARF7 was expressed at a high level, especially in the leaf, flower, and bract tissues (Zhang et al. 2022).

In addition to the regulation by its own promoter, the expression of an *ARF* gene is also affected by the external environment in some cases. For example, incubation in darkness significantly increased *ARF2*, *ARF7*, and *ARF19* expression and decreased *ARF1* expression in *Arabidopsis* leaves (Ellis et al. 2005). Light exposure greatly increased the expression of *ARF8* in *Arabidopsis thaliana* (Tian et al. 2004) but downregulated it in rice (Wang et al. 2007). In the young leaves of blueberry tissue culture seedlings, VcARF1-1, VcARF2-1, VcARF2-5, VcARF6-1, VcARF8-4, VcARF16-1, VcARF18-1 were transcriptionally decreased as pH shifted from 4.8 to 6.8. In contrast, an enhanced expression was observed only at pH 5.2 and 5.8 for VcARF7-1 and VcARF9-1, respectively (Li et al. 2022). For eggplant, all of the SmARFs exhibited rapid responses to salt stress in leaves except SmARF4/5/7B/17/19, in roots except SmARF1A/5/7B, and in stems except SmARF5 (Chen et al. 2022). In addition, application of some exogenous hormones can affect the expression of ARF genes. For example, auxin or ethylene treatment on roots slightly increased the expression levels of *ARF4*, *ARF5*, *ARF16*, and *ARF19* in *Arabidopsis thaliana* and *ARF23* in *Oryza sativa* but slightly decreased the expression levels of *ARF14* and *ARF21* in *Arabidopsis thaliana* and *ARF5* in *Oryza sativa* (Okushima et al. 2005; Li et al. 2006; Wenzel et al. 2007). Exogenous ABA treatment on *Solanum lycopersicum*, the transcription levels of 35 ARF genes were altered, with 28 genes upregulated and 7 genes downregulated (Wei et al. 2022).

6.3 Biological Functions of the ARF Family

The description of the expression pattern of ARFs shows that ARFs are widely expressed in various organs at different developmental stages of each species. Research in the past two decades has discovered that ARFs function throughout the growth and developmental stages of plants and participate in a large number of biological processes related to growth and development, such as embryonic development, leaf extension and senescence, lateral root growth, floral organ development and senescence, fruit development, seed development, and responses to external stresses.

6.3.1 Current Research Status of ARFs in *Arabidopsis thaliana*

Arabidopsis thaliana is regarded as a model plant for scientific research because of its convenience for genetic experiments (Koornneef and Meinke 2010). It has been used in the majority of the studies on the functions of ARF, and even the earliest discovery regarding the functions of *ARFs* was achieved from the phenotype of *arf*-deficient *Arabidopsis* mutants.

Using RNA interfering (RNAi) technology, Ellis et al. constructed multiple *arf1* and *arf2* mutants and found that ARF2 could promote the transitions between multiple growth stages independently of the ethylene and cytokinin response pathways in *Arabidopsis thaliana*. *arf2* mutant plants demonstrated phenotypes related to plant aging, including flowering, rosette leaf senescence, floral organ abscission, and silique ripening. It is worth noting that ARF2 expression increased in senescing leaves. *arf1* phenotypes were not significantly different from wild types, but an *arf1* mutant could upregulate the Aux/IAA gene in plants, consistent with previous findings indicating that ARF1 is a transcriptional repressor. *Arf1 arf2* double mutants enhanced the phenotypes of *arf2* single mutant and had an exaggerated hypocotyl hook, indicating that ARF1 might also participate in regulating floral organ abscission and act in a partially redundant manner with ARF2 (Ellis et al. 2005).

ARF3 and ARF4 contribute mainly to maintaining leaf growth polarity. In plant developmental biology, establishment of polarity is key to organogenesis. Relative to the radially symmetric stem, leaves are bilaterally symmetric organs with polarity along three-dimensional axes, i.e., the proximal–distal, adaxial–abaxial, and medial–lateral axes (Bowman et al. 2002). In *Arabidopsis thaliana*, ASYMMETRIC LEAVES1 (AS1)-deficient mutant *as1* and *as2* cause defects in the leaf adaxial–abaxial polarity, and the leaves are abaxialized filamentous leaves in the various modifier mutant backgrounds. It is widely believed that ARF3 and ARF4, which are regulated by AS1 and AS2, directly act on the leaf growth along the adaxial–abaxial axis (Ishibashi et al. 2013; Iwasaki et al. 2013). The KANADI gene family also regulates the polarized growth of lateral organs, while ARF3 has been shown to change ectopic KANADI activity throughout the plant. Although abnormalities related to *arf3* single mutation are restricted to flowers, *arf3 arf4* double mutants exhibit limited leaf growth on their abaxial sides and showed abnormalities in the polarized development along the dorsal–ventral axis of the gynoecium, resembling the phenotypes of *kan* mutations. In addition, ARF3 and ARF4 are also expressed in the abaxial plane of lateral organs. Therefore, they are crucial to the polarized growth of plants (Pekker et al. 2005). Studies have shown that ARF3 is also involved in the meristem development of lateral roots and flowers in *Arabidopsis thaliana*. ARF3 expression at both the transcriptional and posttranscriptional levels is mediated by *trans*-acting small-interfering RNA (ta-siRNA), and the metabolic pathway of ta-siRNA-ARF is sensitive to auxin and abscisic acid. Yoon et al. (2010) found that ARF3 is involved in the ta-siRNA-ARF metabolic pathway, and its expression can be positively regulated by auxin and negatively regulated by abscisic acid at the posttranscriptional level. This response is the mechanism underlying the role of ARF3 in lateral root development. The *WUSCHEL* (*WUS*) gene plays a key role in flower meristem development. It is positively regulated by *APETALA2* (*AP2*) and negatively regulated by *AGAMOUS* (*AG*). *ARF3* is involved in flower meristem development through negatively regulating *WUS* expression, during which *AG* and *AP2* also play important roles (Yoon et al. 2010; Liu et al. 2014b).

ARF5 is critical for the formation of flowers and radicles. *arf5* mutants show defects in embryo and vascular morphology, and the defects are more remarkable in *arf5 arf7* double mutants, suggesting that *ARF5* and *ARF7* may function redundantly (Hardtke and Berleth 1998). In addition, Harper et al. (2000) confirmed that the effect of *ARF7* varies significantly with the concentration of auxin. *ARF7* regulates the micro process of local cell extension, and its deficiency would result in the disappearance of epicotyl phototropism and hypocotyl gravitropism (Harper et al. 2000). *ARF6* and *ARF8* play an important role in regulating the reproductive growth of flowering plants. *arf6 arf8* double-null mutants have short stamen filaments and undehisced anthers, which results in sterility (Nagpal et al. 2005). Other studies have revealed that *ARF8* is a negative regulator of fruit initiation, and *arf8* mutation causes longer fruit pods and gives rise to parthenocarpic fruit (Goetz et al. 2006). In addition, *ARF8* is also related to petal growth. Compared with the wild type, *arf8* mutants have larger petals with more distinct vein patterning (Varaud et al. 2011).

ARF10 is involved in seed germination and postembryonic development. Liu et al. mutated the recognition site of miR160 to prevent miR60 cleavage of ARF10 in *Arabidopsis thaliana*, and the resultant mutants had developmental defects such as serrated leaves, curled stems, and twisted flowers and siliques. This phenotype did not exist in wild-type and *ARF10*-overexpressing plants (Liu et al. 2007). In *arf10 arf16* double mutants, root cap formation is significantly inhibited, and gravitropism of lateral roots disappears; this phenotype does not appear in *arf10* and *arf16* single mutants (Wang et al. 2005a).

Although ARF10, ARF16, and ARF17 show high evolutionary homology, ARF17 cannot restore the gravitropism in *arf10 arf16* double mutants, suggesting that ARF17 may have different functions from the other two (Wang et al. 2005a). The *arf17* mutant, obtained by Yang et al. (2013) by T-DNA insertion, showed a male-sterile phenotype with normal vegetative growth, indicating that ARF17 plays a key role in pollen development (Yang et al. 2013).

*Arabidopsis arf7 arf*19 double mutants have significantly fewer lateral roots, which does not appear in *arf*7 single mutants. This finding suggests that ARF7 and ARF19 are crucial in lateral root development and have some redundant functions in *Arabidopsis thaliana* (Okushima et al. 2005).

AtARF2 is a transcription factor that can regulate plant senescence. Xue et al. (2022) first screened large-scale genetic reversal mutations and found that ARF2 and PHYTOCHROME-INTERACTING FACTOR 5 (PIF5) are a key transcriptional factor responsible for transcriptional regulation in the ABS3-mediated aging pathway. Further studies found that ARF2 and PIF5 interact directly and jointly regulate the target genes in the aging pathway, such as the key aging promoting factors ORESARA1 (ORE1) and STAY-GREEN1 (SGR1). At the same time, RNA-Seq and genetic analysis showed that ARF2 and PIF5 jointly controlled the transcriptional regulation of senescence downstream of ABS3 in an interdependent and independent manner. These data establish that ARF2 and PIF5 are required for the ABS3-mediated senescence (Xue et al. 2022).

6.3.2 Current Research Status of ARFs in Rice (*Oryza Sativa*)

OsARF1 is the first *ARF* gene cloned from monocot plants. Attia et al. (2009) found that the expression level of *OsARF1* was significantly higher in embryonic tissues than in vegetative tissues, suggesting an essential role of ARF1 in embryo development. Plants with RNAi-silenced *OsARF1* expression showed extremely slow growth and curled leaves; these plants tillered but were sterile (Attia et al. 2009).

It has been confirmed that ARF12 and ARF25 are involved in the process of phosphate homeostasis in *Oryza sativa*. In an acid phosphatase (APase) activity assay, Wang et al. (2014) found that the phosphorus content and the expression of genes related to phosphate transport increased significantly in both *arf12* single mutants and *arf12 arf25* double mutants (Wang et al. 2014). OsARF16 also contributes to the phosphate metabolism in the plant body. Research has shown that ARF16 is involved in the inhibitory effect of cytokinin on phosphate transport from roots to shoots. In *arf16*-deficient mutants, the phosphate concentration increased significantly (Shen et al. 2014).

Floral organ development is critical to plant reproduction and seed quality, and ARF19 is involved in this developmental process in *Oryza sativa*. A rice *osarf19* null mutant obtained by Zhang et al. (2016) using T-DNA insertion displayed three types of obvious floral abnormalities: an enlarged palea with a curled tip that affects flower closure, palea degeneration, and an additional lemma. Cell morphological analysis showed that floral organ abnormalities were caused by abnormal cell morphology, and the cells were significantly elongated in *osarf19* mutants (Zhang et al. 2016). OsARF19 overexpression can cause an enlarged lamina inclination due to increased adaxial cell division. Zhang et al. (2015a) attributed this effect to the regulatory effect of ARF19U on OsGH3-5 and brassinosteroid insensitive 1 (OsBRI1) by using chromatin immunoprecipitation (ChIP) and yeast one-hybrid technology (Zhang et al. 2015a).

Rice morphology determinant (RMD), an actin-binding protein, can regulate the auxin–actin feedback loop and directly affect the cell growth and morphology of rice plants. Auxin promotes the expression of RMD by mediating the binding of OsARF23 and OsARF24 to the RMD promoter, through which auxin triggers the structural changes in F-actin. RMD-dependent F-actin can, in turn, affect auxin signal transduction in cells, including polarized auxin transport, localization and recycling of auxin efflux carriers, and auxin distribution in root cells (Li et al. 2014).

Previous study had revealed that a trans-acting small interfering RNA targeting AUXIN RESPONSE FACTORS (tasiR-ARF) regulates lemma polarity establishment via posttranscriptional repression of AUXIN RESPONSE FACTORS (ARFs) in rice. The transgenic plants with ectopic OsARF3a expression showed irregular phyllotaxis with rolled-up or threadlike leaves, indicating the essential roles of proper OsARF3a expression in abaxial–adaxial polarity establishment (Song et al. 2012). TasiR-ARF formation requires RNA-DEPENDENT RNA POLYMERASE 6 (RDR6). In osrdr6-1 mutants, the lemma is abaxialized owing to disruption of OsRDR6-dependent tasiR-ARF accumulation in adaxial cells, which leads to ectopic expression of ARF3a. Si et al. (2022) identified a suppressors, hf1, by genetic screening for suppressors of the lemma defect mutant osrdr6-1. HF1 encoded a MYB transcription, OsKANADI1, which could directly bind to the intron of OsARF3a in vitro and in vivo and promote OsARF3a expression at the transcriptional level. Thus OsKANADI1 and tasiR-ARFs synergistically maintain the proper expression of OsARF3a and contribute to rice lemma development (Si et al. 2022).

In rice, ARF genes also regulate the host's immune defense against virus. By using the rice-rice dwarf virus (RDV) model system, Qin et al. (2020) reveal a novel auxin-IAA10-ARFs-mediated signaling mechanism employed by rice and RDV for defense and counterdefense responses. RDV infection triggered the increase of auxin biosynthesis and accumulation in rice, while exogenous auxin treatment decreased the level of OsIAA10 protein, thus releasing a group of OsARF interacting with OsIAA10 to mediate downstream antiviral response. The functional loss mutants of osarf12 or osarf16 showed decreased resistance to RDV, while the osarf11 mutants showed increased resistance to RDV. In turn, OsARF12 activates downstream OsWRKY13 expression by directly binding to its promoter, and the dysfunctional mutant of oswrky13 shows reduced resistance. These results showed that OsARF11, 12, and 16 differently regulated the antivirus defense ability of rice (Qin et al. 2020).

6.3.3 Current Research Status of ARFs in Tomato (*Solanum lycopersicum*)

Tomato is a model plant for research on fleshy fruit ripening (Klee and Giovannoni 2011). The existing studies on the ARF family in tomato are mostly focused on fruit development and ripening, and the details will be discussed in the following section.

6.4 Mechanism of ARF Function

The mechanism underlying ARF action is that, through binding to auxin early response factors, ARF participates in the regulation of auxin gene expression in the signal transduction pathway, thereby regulating plant growth and development (Tatematsu et al. 2004). The structure of a gene determines its function. ARF genes and Aux/IAA have the same structural domains, namely, domain III and domain IV of CTD, leading to a close relationship between the two in the regulation of plant growth (Guilfoyle et al. 1998). Therefore, the mechanism underlying the binding of ARF to Aux/IAA has always been a research hotspot on ARF–Aux/IAA interaction.

6.4.1 ARF–Aux/IAA Binding Mechanism

In recent years, studies have shown that ARF- and Aux/IAA-deficient mutants are insensitive to exogenous auxin, and ARF and Aux/IAA proteins may be the center of sensing the auxin concentration gradient and converting the information into gene expression changes, thereby altering plant morphology and developmental patterns. A large number of studies using genetic and molecular/biochemical experimental methods have confirmed that ARF and Aux/IAA proteins are transcription factors of primary auxin response genes, and there are specific ARF–ARF and ARF–Aux/IAA interactions. Guilfoyle and Hagen proposed a model to explain how ARF and Aux/IAA regulate the expression of early auxin response genes. In this model, regardless of whether the transcription factor ARF binds to the target TGTCTC element or not, Aux/IAA repressors are thought to dimerize with ARF activators via their domains III and IV under conditions in which auxin concentrations are low, resulting in repression of auxin response genes. Exogenous Aux/IAA protein may directly bind to endogenous ARF that has been bound to the target sequence and therefore affect the ability of ARF to activate AuxREs. It is also possible that exogenous Aux/IAA might bind to non-DNA-bound ARF and therefore stop its activation function by preventing its interactions with AuxRE-bound ARF (Guilfoyle and Hagen 2007).

Another model proposed that the ARF binds AuxREs regardless of the auxin concentration in cells or the auxin response factors being repressed or activated. In addition, auxin may affect the ARF–ARF interactions that are essential for transcriptional activation or repression. Under conditions where auxin concentrations are low, ARF binds to AuxREs and leads to repression by combining the action of the transcriptional repressors that have not bound to the target sequence. This repression may be subject to the impact of Aux/IAA protein and inactive ARF protein, which both contain a CTD and can interact with each other. When the auxin concentration increases, the Aux/IAA repressors can be destroyed by the ubiquitin-proteasome pathway, and the destruction causes the transcriptional inactivation of ARF bound by the target DNA. The resultant activation function can be strengthened by the binding of Aux/IAA with ARF or Aux/IAA activators that have not bound to the target sequence (Ulmasov et al. 1999).

6.4.2 Mechanism for the Role of ARFs in Auxin Signal Transduction

The auxin signal transduction pathway includes the processes of signal recognition, initiation of downstream auxin-related gene expression, and physiological responses of plants. The main regulation model of auxin signal transduction involves the coordination of ARF, Aux/IAA, and the ubiquitin-mediated protein degradation pathway (the Skp1–Cul1–F-box-protein (SCF) system). In the negative regulation model that is broadly recognized, auxin first binds to specific receptors or carriers and transmits the signal to Aux/IAA, leading to conformation changes of the key proline residues on Aux/IAA. The changes affect SCF through E1 or other SCF system components, triggering Aux/IAA ubiquitination. Next, the modified protein is recognized and degraded by the 26S protease. The decrease and degradation of Aux/IAA reduce the abundance of Aux/IAA homologous and heterogeneous dimers and Aux/IAA–ARF dimers but increases that of ARF–ARF dimers. These dimers can bind AuxREs of auxin-inducible factors to enhance the efficiency of transcriptional expression (Ulmasov et al. 1999).

6.5 Research Progress on the Function of ARF Family Transcription Factors in Plants

Auxin, the first discovered family of plant hormones, plays a key role in all stages of plant growth and development. As transcription factors, ARFs are critical in auxin signal transduction. ARF proteins can regulate gene expression by binding to AuxREs in the promoter region of auxin response genes. During auxin signal transduction, *Aux/IAA*-encoded proteins regulate gene expression by binding to *ARF* genes and inhibiting their effect. The mRNA expression of *IAA*, *GH3*, and *SAUR* genes can be rapidly induced by auxin, and therefore, they were named primary auxin response genes. During the flowering and fruiting period of tomato, adverse temperatures (either low or high) may cause poor pollination and fertilization in addition to flower and fruit abscission. These responses directly decrease the rate of fruit set and increase the number of abnormal fruits, seriously affecting the yield and quality of tomato. Previous studies have shown that auxin-related genes, especially *ARF* and *Aux/IAA* genes, are vital in parthenocarpy in tomato.

In recent years, advancements in plant genome sequencing have made it possible to deepen whole-genome studies on the *ARF* gene family. A series of studies have been conducted to investigate the expression characteristics, classification, biological functions, and regulatory mechanisms of *ARF* genes. For example, in tomato, *ARF3*, *ARF5*, *ARF6*, *ARF9*, *ARF10*, *ARF13*, *ARF15*, and *ARF17* have similar functions, and their expression levels peak during the period from young fruit to fruit ripening, while *ARF1*, *ARF9*, *ARF11*, *ARF15*, and *ARF16* are obviously expressed in the flower developmental stage, and in particular, the expression of *ARF16* is the highest in the flower buds (Kumar et al. 2011).

Interfering with small RNA production is a common strategy used to protect plants from viral infection. A unique class of small RNAs that require microRNA and short interfering (siRNA) biogenesis for their synthesis is named *trans*-acting small interfering RNAs (ta-siRNAs). Tomato (*Solanum lycopersicum*) mutants with filamentous leaves represent a class of phenotype that mimics viral infection symptoms, including shoestring leaves with weakened leaf blade expansion. Four *WIRY* genes are confirmed to be involved in siRNA biosynthesis, and in their corresponding mutants, ta-siRNAs that regulate *ARF3* and *ARF4* decrease. Reducing *ARF3* and *ARF4* expression can rescue the shoestring leaf lamina, while increasing the expression of either *ARF3* or *ARF4* can phenocopy shoestring leaves. Overexpression of these *ARFs* in *Arabidopsis thaliana*, tobacco (*Nicotiana tabacum*), and potato (*Solanum tuberosum*) does not result in the wiry leaf phenotype, suggesting that ta-siRNA regulates the expression of *ARFs*. This effect suggests that ta-siRNA enrichment may serve as a universal sensor for interference with small RNA biogenesis (Yifhar et al. 2012).

Increasing evidence has demonstrated that ARFs are crucial to plant development, including leaf extension and senescence, lateral root development, determination of floral meristem identity, flower abscission, fruit set and development, embryo formation, and responses to environmental stimuli (Ckurshumova et al. 2014; Guan et al. 2014; Liu et al. 2014b). Goetz et al. (2007) found that abnormal expression of ARF8 promoted the formation of parthenogenesis in tomato and *Arabidopsis*. Later, researchers found that downregulation of *ARF* genes can induce changes in pectin and tissue structure in tomato. In 2009, an anatomic analysis of tomato ovaries showed that the *SlARF7* transcript level significantly increased during flower development and remained at a constant high level in mature flowers but declined within 48 hours after pollination. Decreasing *SlARF7* mRNA levels in tomato plants resulted in seedless (parthenocarpic) fruits, which were heart-shaped and had a thicker pericarp. These results suggest that, in tomato, *SlARF7* acts as a negative regulator of fruit set prior to pollination and fertilization, and it regulates the auxin response during fruit growth (de Jong et al. 2009). Subsequently, new findings indicated that *ARF9*, as a regulator, regulates fruit set and development in tomato (de Jong et al. 2015). In 2011, researchers found that ARF12 in rice plants could regulate root elongation and affect iron accumulation

(Qi et al. 2012). In the same year, researchers discovered 21 candidate functional *ARF* genes from the tomato genome and obtained the full-length cDNA sequences of 15 *ARF* genes by sequencing polymerase chain reaction (PCR) products (Wu et al. 2011). These findings are of great value to the functional research of tomato *ARF* genes. In 2012, it was also reported that derepression of *ARF10* could inhibit leaf growth in tomato (Hendelman et al. 2012). In addition, *ARFs* are crucial to the crosstalk between auxin and other hormones such as GA, ABA, brassinosteroid, and ethylene signaling (Zouine et al. 2014). A study showed that methyl jasmonate (MeJA) treatment inhibited the growth process of apple calli, and ARF8 played a negative regulatory role in this pathway. The results of ectopic expression in Arabidopsis showed that MdARF8 could reduce the sensitivity of Arabidopsis to MeJA and alleviate the phenotype of promoting leaf senescence and inhibiting taproot elongation (Qiao et al. 2022).

6.5.1 Research Progress on the Role of ARF Family Transcription Factors in Tomato Fruit Development and Ripening

Tomato is an important vegetable crop produced worldwide. China's annual production of tomato accounts for more than 20% of the world's total output. As a hormonal factor, auxin plays an important role in the regulation of fruit set and development in tomato (Hendelman et al. 2012). The level of auxin increases rapidly in the tomato ovaries two days after pollination and fertilization (Mariotti et al. 2011). Auxin treatment enables unfertilized ovaries to form fruit without fertilization. Similarly, overexpression of genes related to auxin synthesis or response, such as *iaaM* and *rolB*, affects the level of auxin, and overexpression of ARF receptor protein TIR1 can also lead to parthenocarpy in tomato (Ren et al. 2011). Among the ARF family transcription factors, ARFs 5–8 and 19 are transcription activators and ARFs 1–4, 9–18, and 20–22 are transcription repressors (Tiwari et al. 2003). In a 2002 study, researchers found that a transcription factor of the ARF family, DR12, plays an important role in determining the color of fruit pericarp. Using a DR12–GFP gene fusion construct for localization, the study confirmed that DR12 is located in the nucleus, indicating that DR12 is indeed a transcription factor. In addition, transgenic tomato fruits with either perturbed DR12 expression or cosuppressed DR12 expression displayed a pleiotropic phenotype that included dark-green immature fruit, unusual cell division in the fruit pericarp, blotchy ripening, enhanced fruit firmness, upward-curling leaves, and increased hypocotyl and seed growth (Jones et al. 2002). The phenotype might be attributed to a perturbation of auxin response signaling, whereas the expression of genes related to auxin regulation remained unchanged in the plants (Jones et al. 2002). Subsequent studies showed that the emergence of the phenotype with dark-green immature fruit in *ARF4* transgenic tomato could be explained by the accumulation of chloroplasts regulated by *ARF4* transcription factor, especially in the immature fruit. Moreover, transgenic tomato with ARF4 downregulation had a lower starch content in its early fruit compared to that in wild-type plants during the same stage. The chlorophyll content and photosynthesis were also enhanced in transgenic plants, which may explain why enhanced photosynthesis leads to the increased starch content in transgenic tomato fruits. The expression of ARF4 is enhanced in the pericarp; however, in the early stage of fruit ripening, ARF4 expression decreases abruptly, accompanied by an increase in the starch content. Gene expression analysis in transgenic plants revealed that the increased starch content in transgenic fruit is caused by the upregulation of genes and enzymes related to starch synthesis. These results suggest that ARF4 is a negative regulator. In summary, the ARF family transcription factors can regulate the starch content in fruit during fruit ripening. This conclusion for the first time links the function of ARF transcription factors with auxin signaling, chlorophyll activity, and glucose metabolism.

In 2009, Jong et al. found a tomato *ARF* gene homologous to *Arabidopsis NPH4/ARF7* and named it *Solanum lycopersicum ARF7* (*SlARF7*), which is highly expressed in unfertilized mature ovules. Further analysis showed that *SlARF7* expression remained high during flower development until flower maturation, but it was downregulated 48 hours after pollination fertilization. Transgenic plants with suppressed *ARF7* expression could form seedless (parthenocarpic) fruits, which were heart-shaped and had a rather thick pericarp due to increased cell expansion. Taken together, these findings show that the *ARF7* gene negatively regulates fruit set until pollination is completed, and it also regulates auxin production during fruit growth in tomato. *ARF3*, another ARF family member in tomato, encodes a protein that contains both B3 and ARF functional domains but lacks the AUX domain. Gene expression analysis showed that the expression level of *ARF3* was particularly high in epidermal cells and trichomes, and an in situ hybridization assay revealed that *ARF3* was also expressed in guard cells of leaf epidermis. Transgenic plants with downregulated *ARF3* expression have a less dense distribution of epidermal cells in their leaves and significantly fewer type I, V, and VI trichomes, which indicates that *ARF3* is critical to the formation of trichomes and epidermal cells. Moreover, the number of xylem cells decreases significantly in transgenic plants with *ARF3* downregulation. In addition, 51 differentially expressed genes (DEGs) belonging to 14 transcription factor families, such as *MYB*, *bHLH*, *WD40*, and *C2H2* zinc finger, have been found and identified by transcriptome sequencing. Another 27 DEGs are involved in metabolism and signal transduction pathways of phytohormones, such as auxin, ethylene, and gibberellin. These results indicate the important role of transcription factors and auxin-dependent hormonal regulation of trichome formation in tomato, which also reveals the special functions of the ARF family of transcription factors in fruit development.

In 2015, researchers found that *ARF9* was specifically expressed during fruit set and early fruit development. *ARF9* is expressed in response to auxin. The mRNA level of *ARF9* is relatively high in ovules, embryos, and pericarp, while other tissues, such as axillary and root meristems, may also have a high level of cell division activity. Transgenic tomato plants with *ARF9* overexpression may have weakened cell division activity in fruits, which makes fruits smaller than those of wild-type tomato plants. Conversely, transgenic plants with downregulated *ARF9* expression have significantly larger fruits than those of wild-type tomato plants. These findings suggest that *ARF9* regulates early fruit development by negatively regulating cell division (de Jong et al. 2015).

Using quantitative PCR assays for expression analysis, Zhang et al. (2015a) analyzed *ARF* genes involved in trichome formation and development, and they found that ARF3, 4, 6, and 10 expression were relatively high in trichomes. The role of *ARF* genes in trichome formation and development was explored by the RNA interference (RNAi) transgenic approach. It was found that plants with *ARF3* downregulation had thinner leaves, decreased density of epidermal cells, and fewer xylem cells (Zhang et al., 2015b). The numbers of type I, V, and VI trichomes in leaves were significantly smaller in the RNAi transgenic plants than those of wild-type plants, indicating that *ARF3* is involved in formation of trichomes in tomato. In addition, the molecular regulatory pathway of trichomes showed certain similarity between tomato plants and *Arabidopsis thaliana* (Zhang et al. 2015a).

The involvement of ethylene in fruit ripening has been relatively thoroughly studied, but very few studies have investigated the regulation of fruit ripening by the interactions between ethylene and other hormones. It has been found that *ARF2A* plays an important role in fruit ripening. *ARF2A* is specifically expressed in fruit, and its expression level significantly decreases in *rin*, *nor*, and *nr* mutants. *ARF2A* can also respond to endogenous ethylene, auxin, and abscisic acid. In *ARF2A*-overexpressing plants, the fruit displayed a mottled pattern of ripening, with some spots maturing and turning red earlier at an accelerating rate (Hao et al. 2015). In transgenic tomato overexpressing *ARF2A*, fruit ripening was delayed due to early release of ethylene and suppression of the ethylene signal transduction pathway, suggesting that the *ARF2A* gene is ethylene dependent. Components of ethylene signaling and responses could be induced in both green and red regions of fruits, which could regulate fruit ripening (Hao et al. 2015). Hormone analysis revealed that changing the expression level of *ARF2A* significantly affected the contents of abscisic acid, cytokinin, and salicylic acid but not gibberellin and auxin metabolites. Silencing *ARF2A* generated plant phenotypes that displayed delayed maturation, parthenocarpy, and hormone disorder. Finally, it was found that the ARF2A protein can form both homodimers and heterodimers with the ASR1 protein, suggesting that the ASR1 protein may be associated with ABA and ethylene dependent. The findings above reveal that ARF2A interacts with ethylene signaling molecules and other hormones to regulate fruit tissue volume to regulate the process of fruit ripening.

ARF2 has been reported to be a pleiotropic developmental regulator in *Arabidopsis* and tomato. According to the results of existing studies, ARF2 is expressed in all tomato tissues, and its expression level is the highest in flowers. During flowering, *ARF2* shows dynamic expression changes in the calyx, stamens, ovary, and petals. In hormone treatment experiments, the expression of *ARF2* was upregulated by auxin and gibberellin treatment on tomato buds but downregulated by ethylene treatment. In transgenic tomato plants with upregulated *ARF2* expression, the analysis of phenotypes and physiological parameters showed that *ARF2* regulated lateral root formation and flower organ senescence. These phenotypes were believed to be caused by the expression changes in ARF and ethylene response factor (ERF)-related genes. The research findings above expand the knowledge regarding the function of *ARF2* in plants and the mechanism underlying the regulatory role of auxin in plant growth and development (Rena et al. 2017).

Auxin generally regulates plant growth and development by regulating gene expression, for which the transcription factors in the ARF family bind to the promoter regions of target genes to activate or repress the expression of downstream genes. In *Arabidopsis thaliana*, *ARF6* and *ARF8* have been demonstrated to promote inflorescence stem elongation and late stages of petal, stamen, and gynoecium development. All seed plants studied thus far have *ARF6* and *ARF8* orthologs and orthologous sequences of miR167, which targets *ARF6* and *ARF8*. However, it remains unknown whether the functions of these genes in flower development are conserved. To answer this question, the expression level of MIR167a in transgenic tomato plants with downregulated *ARF6* and *ARF8* was analyzed. Plants overexpressing MIR167a had smaller leaves, shortened internode length, shortened petal length, and smaller stamens. More importantly, the transgenic plants were female sterile because wild-type pollen failed to germinate on the stigma surface. Transcriptome analysis of transgenic and wild-type tomato identified many genes with significantly altered their expression levels. Gene Ontology (GO) analysis revealed that these DEGs encode products with functions in "transcription regulation," "cell wall," and "lipid metabolism" categories. Orthologs of these genes were also differentially expressed in *Arabidopsis arf6 arf8* mutants. The results above indicate that *ARF6* and *ARF8* play conserved roles in regulating the growth and development of vegetative and flower organs in dicots (Liu et al. 2014a, 2014b).

Hendelman et al. (2012) found and identified the orthologous *ARF10* (*SlARF10*) in tomato. They demonstrated that miR160 posttranscriptionally regulates *ARF10* and plays an important role in tomato development. In wild-type tomato plants, *SlARF10* is primarily expressed in the pericarp of mature and red ripened fruit;

this expression profile is opposite to that of miR160 in tomato fruit. Overexpressing *SlARF10* did not alter tomato development in wild-type plants, but transgenic tomato plants with suppressed Sl-miR160 expression had narrow leaflets, sepals, and petals and abnormally shaped fruit. During compound leaf development, *SlARF10* accumulation specifically inhibited leaflet blade outgrowth but did not affect other auxin-dependent processes such as leaflet initiation and formation. In addition, blade size was negatively correlated with *SlARF10* accumulation, further indicating that the SlARF10 protein is a transcriptional repressor of leaflet lamina outgrowth. The researchers then proposed that repression of SlARF10 transcript by Sl-miR160 is essential for blade outgrowth and early fruit development in tomato (Hendelman et al. 2012). And also, Yuan et al. (2018) found that SlARF10 was highly expressed in green fruit. Overexpression of SlARF10 in fruit produced a dark-green phenotype while knockdown by RNAi produced a light-green phenotype. Autofluorescence and chlorophyll content analysis confirmed this phenotype, indicating that SlARF10 plays an important role in chlorophyll accumulation. The regulation of SlARF10 expression had a positive effect on photosynthesis of tomato leaves and fruits. In addition, the accumulation of starch, fructose, and sucrose in the fruits of the lines overexpressing SlARF10 increased, while the accumulation of starch and sucrose in the fruits of the lines that inhibited the expression of SlARF10 decreased. The regulation of SlARF10 expression changes the expression of starch biosynthesis gene AGPases. SlARF10 is regulating the expression of SlGLK1, POR, CBP1, and CBP2, which are related to chlorophyll metabolism and regulation. These results indicate that SlARF10 is involved in chlorophyll accumulation by transcriptional activation of SlGLK1 expression in tomato fruit (Yuan et al. 2018).

In 2018, Liu et al. found that SlARF regulates fruit set and development via the mediation of auxin and gibberellin signaling. SlARF5 was expressed at high level in castrated ovary, but at low level in pollinated ovary. miRNA-SLARF5 line showed ovary growth and formed seedless fruit after castration. Compared with wild-type fruit, transgenic parthenocarpy fruit has less chamber tissue, smaller fruit size and weight. RNA-Seq analysis demonstrated that several genes involved in the auxin-signaling pathway were downregulated, whereas some genes involved in the gibberellin-signaling pathway were enhanced by the decreased SlARF5 mRNA levels in transgenic plants, which indicated SlARF5 is an essential component of auxin signaling during tomato fruit development (Liu et al. 2018).

ARF6A localized in the nucleus is a transcriptional activator of tomato fruit development. Compared with wild-type (WT) plants, overexpression of SLARF6A increased chlorophyll content in tomato fruits and leaves, while downregulation of SLARF6A decreased chlorophyll content. The overexpression of SLARF6A gene increased the rate of photosynthesis and the accumulation of starch and soluble sugar in tomato leaves and fruits, while the silencing of SLARF6A gene led to the opposite phenotype of tomato leaves and fruits. RNA sequence analysis showed that the regulation of SLARF6A expression changed the expression of genes involved in chlorophyll metabolism, photosynthesis, and glucose metabolism. SlARF6A directly binds to the promoters of SlGLK1, CAB, and RBC genes, and positively regulates the expression of these genes. And also, overexpression of SLARF6A also inhibited fruit ripening and ethylene production, while downregulation of SLARF6A promoted fruit ripening and ethylene production. This may be due to SLARF6A negatively regulates the expression of SAMS1, a key gene in ethylene synthesis (Yuan et al. 2019).

In 2022, Xu et al. found that SlARF4 participates in the molecular mechanism of auxin and abscisic acid antagonistic regulation of ascorbic acid (AsA) production in tomato. By measuring the contents of natural auxin IAA and AsA in tomato pericarp, it was found that there was a negative correlation between auxin and AsA content during fruit development and ripening. Then treated with different concentrations of IAA, it was found that IAA inhibited the accumulation of AsA in tomato. By analyzing the gene expression data of tomato pericarp treated with IAA, data showed that ARF family genes were upregulated, among which SIARF4 induced expression was the strongest, and its expression pattern was consistent with the change pattern of auxin concentration, but contrary to that of AsA, indicating that SIARF4 may be involved in AsA biosynthesis during fruit development and ripening. And then the SIARF4 overexpression and knockout lines proved that the inhibition of AsA accumulation induced by auxin was indeed mediated by SIARF4. The metabolic pathway of SIARF4 is affected in lines with downregulated AsA expression. Specifically, SlARF4 inhibits the transcription factor gene SlMYB11 in transcription, thus regulating the accumulation of AsA by regulating the transcription of AsA biosynthesis genes L-galactose-1-phosphatase, L-galactose-1 Lactone dehydrogenase, and dehydroascorbic acid in tomato fruit (Xu et al. 2022).

In plants, miRNAs play important roles in regulating auxin signal transduction by inversely regulating the expression of ARFs. ARF10/16/17, targeted by miR160, are indispensable for plant development, but very few studies have explored their functions in tomato (*Solanum lycopersicum*). Some researchers used short tandem target mimics (STTM) to suppress miR160 expression for assessment of its roles in tomato development. A Northern blot assay showed that miR160 was abundantly expressed in the developing ovary. Compared with wild-type plants, the plants with downregulated miR160 expression demonstrated the excessive elongation of the ends of ovaries and thinning of the placenta. After fertilization, these morphological changes led to the formation of elongated, pear-shaped fruits. In addition, STTM160-expressing plants displayed abnormal floral organ abscission, accompanied by smaller numbers of leaves and sepals, and petals that easily dehisced. These findings indicate that miR160 plays a pivotal role in the auxin-mediated fruit developmental processes. In addition, miR160 mutation was always associated with the upregulation of *ARF10A*, *ARF10B*, and *ARF17*, of which *ARF10A* upregulation was the most prominent. Although *ARF16A* is believed to be a target of miR160, the expression of *ARF16A* did not change significantly in response to miR160 downregulation in transgenic tomato, suggesting that *ARF16A* may be regulated by cleavage of other miRNAs. Previous studies have suggested that *ARF10A* and *ARF17* were the negative regulators of the ARF family. Therefore, it is believed that adjusting the expression of the targets of miR160 (ARF10A is the main target) can regulate both auxin-mediated ovary patterning and floral organ abscission and lateral organ lamina outgrowth.

6.5.2 Research Progress of ARF Family Transcription Factors in Fruits of Other Species

Melon, a crop belonging to the family Cucurbitaceae, is an important fruit worldwide and has high economic value.

To determine the whole-genome characteristics of the ARF family in melon, Wu et al. (2020) identified ARF based on the information extracted from the melon genome database and analyzed the expression of ARF during early development of melon by qRT-PCR. From the melon genome, they identified 17 genes encoding ARF proteins. The expression of these ARFs varied in different tissues; for example, ARF16, ARF18, and ARF19 were highly expressed in fruit. During the early development of melon fruit, the expression of ARF9, ARF16, ARF19, ARF1, ARF2, ARF3, and ARF5 differed in response to the different treatments, suggesting that they may be related to early melon development (Wu et al. 2020).

China, as one of the main countries of origin for sweet orange (*Citrus sinensis*), has a wide range of excellent varieties and rich species resources. In South China, sweet orange is a main fruit species, and this region is the primary producer of this species worldwide. In recent years, with the completion of whole-genome sequencing of different species such as poplar, tomato, corn, soybean, and banana, the whole-genome database of sweet orange has also been completed (Xu et al. 2013). Research has confirmed that sweet orange has 19 *ARF* genes (*CiARFs*). Among them, *CiARF3/4* belong to Class I, the same as *AtARF*3/4 in *Arabidopsis thaliana*, as they are all involved in the regulation of flower organ formation and development timing and patterning. *CiARF1/2/9/12/18* can be grouped into Class II, the same as *AtARF1/2/9/18*, which are involved in regulating leaf senescence, flower organ abscission, and auxin homeostasis. *CiARF5/6/7/8/10/19* and *AtARF5/6/7/8/19* all belong to Class III and regulate auxin-mediated flower development, morphogenesis, lateral root formation, and other processes. The citrus *ARF* gene family has been characterized more comprehensively through phylogenetic analysis, chromosome localization, and analysis of conserved protein-coding domain and cis-elements in promoter regions of genes. Quantitative PCR analysis showed that the expression patterns of *CiARFs* varied among different tissues and developmental stages in sweet orange. Some *CiARF* genes demonstrated organ/tissue-specific expression patterns: they were highly expressed in vegetative organs and relatively less expressed in reproductive organs. In experiments in which IAA and naphthyl anthranilic acid (NPA) were used to treat callus samples, the expression of *ARF1/3/4/6/8/15/16/17* decreased after six hours of treatment and then rebounded due to the influence of local high-concentration IAA. NPA treatment at concentrations of 5 μmol/L and 10 μmol/L had a strong inhibitory effect on ARFs, demonstrating that different ARF genes respond differently to different concentrations of IAA and NPA. In transgenic *Arabidopsis thaliana* plants with *ARF16* overexpression, the seeds germinated earlier, and the seedlings had longer roots when compared with those of the wild type, which confirms that ARF16 enhances seed germination and promotes root growth (Li et al. 2015).

Eggplant (*Solanum melongena* L.) is widely cultivated all over the world and is one of the main vegetable crops favored by Chinese urban and rural residents. Breeding seedless or less-seed eggplant varieties is a goal and research hotspot pursued by breeders in China and other countries. At present, exogenous hormone application is the major way to obtain seedless fruit in production, which requires high labor and production costs and might also bring some negative effects due to hormone use. Therefore, it is of great practical value to study the mechanism underlying the function of hormones in parthenocarpy and develop genetic engineering technology to create seedless eggplants. Various studies have shown that the formation of seedless (parthenocarpic) fruit is regulated by various plant hormones such as auxin and GA, indicating that their response genes may be involved in the regulation of parthenocarpy. *ARF* genes are regulators of the primary auxin response genes in plants. The full-length sequence of the eggplant *ARF8* gene with a complete reading frame and poly(A) tail structure has been obtained, and a complete gene sequence of eggplant has been cloned by RT-PCR and rapid amplification of cDNA ends (RACE) technology. Members of the ARF transcription factor family have conserved domains and very distinct sequence characteristics. The newly cloned putative eggplant *ARF8* (*SmARF8*) has all the sequence characteristics, such as DBD, CTD, and nonconserved domains, possessed by ARF homologs, indicating that *ARF8* is an *ARF* homolog in eggplant. After exogenous application of auxin for 5–10 minutes, the expression of various Aux/IAAs increased rapidly, indicating that they are the primary response genes of auxin. Further amino acid sequence similarity analysis showed that *SmARF8* had a gene identity of 54.7% with *AtARF8*, which regulates parthenocarpy-related genes in *Arabidopsis thaliana*, and 54.4% with *SlARF8*, which was suspected to regulate parthenocarpy-related genes in tomato. Phylogenetic analysis showed that putative *SmARF8* had a close phylogenetic relationship with these two genes. Compared with other *AtARFs*, the putative *SmARF8* was phylogenetically closest to *AtARF8*. The results indicated that the cloned *SmARF8* gene might be an *ARF* gene family member regulating parthenocarpy in eggplant. The same regulation pattern in *Arabidopsis thaliana* and tomato, in which *ARF* gene regulates parthenocarpy, might also be present in eggplant. The expression of *ARF8* was significantly reduced in newly formed fruits in *LER* wild-type plants of *Arabidopsis thaliana* but was found to be significantly elevated in 5-mm tomato fruits, because fertilization can change the transcriptional expression level of *ARF8* genes. *ARF8* expression decreased in fertilized flowers after fully blooming and significantly increased after fruit formation. It is also possible that *ARF8* expression is reduced by the fertilization process. The expression pattern of the *ARF8* gene was basically the same in tomato and eggplant, significantly differing from that of *Arabidopsis thaliana*, possibly because tomato and eggplant both belong to the family Solanaceae and have closer genetic relationship, and the regulatory expression patterns of homologous genes evolved from the same ancestor are more similar to each other (Du et al. 2016).

Banana, an important economic fruit cultivated in South China, usually undergoes dramatic physiological and biochemical changes during ripening process. microRNAs have emerged as critical regulators in diverse aspects of plant growth and development. Among the plant hormone response pathways, the relationship between auxin and miRNA has been most studied. Kong et al. (2022) proposed a mode of action of miRNA-mediated auxin signaling in ethylene-induced banana fruit ripening. Combined with bioinformatic prediction, degradome data validation and transient expression assays in tobacco leaves, miR160 was found specifically targeted ARF12 and miR167 specifically targeted ARF7. When the banana fruit is treated with exogenous ethylene, fruit ripening was induced and this process was coupled to an elevated level of auxin in fruit and enhanced auxin signaling. At the same time, both miRNAs were suppressed by ethylene treatment with a descending pattern in expression, while the expression level of ARF12 and ARF7 targeted respectively by miR160 and miR167 maintained much higher in the ethylene-treated pulp compared with the control. A good anti-correlation between miRNA accumulation and their target expression indicated that auxin signaling pathway has been enhanced by ethylene treatment via suppressing the miRNA effect in the banana pulp. In addition, miR160 and miR167 tended to target all members of Class IV and a subgroup of Class I ARF family, respectively, which means the miRNA-ARF module may play potential role in more physiological processes (Kong et al. 2022). Moreover, MaARF2 and MaARF12/MaARF24 can bind to starch-branching enzyme MaSBE2.3 and upregulate and downregulate its expression respectively, thus controlling starch accumulation and amylopectin synthesis in banana fruit (Miao et al. 2020).

The plant hormone auxin plays an important role in various aspects of plant development, such as cell extension, division, and differentiation, as well as in organ and tissue development, and tropism. Auxin response factors (ARFs) are involved in auxin-mediated transcriptional regulation in plants. More and more evidence shows that ARF genes are indispensable in the process of plant fruit development and ripening, and it often interacts with other hormones, such as asethylene, abscisic acid, brassinosteroid, and jasmonic acid. In-depth studies of the molecular functions of ARFs in fruit development not only generate fundamental insights into the common mechanisms that underlie fruit development in plants but also provide the potential to improve the production and quality of horticultural plants.

References

Attia, K.A., Abdelkhalik, A.F., Ammar, M.H. et al. (2009). Antisense phenotypes reveal a functional expression of OsARF1, an auxin response factor, in transgenic rice. *Curr Issues Mol Biol* 111: I29–I34.

Benjamins, R. and Scheres, B. (2008). Auxin: the looping star in plant development. *Annu Rev Plant Biol* 59: 443–465.

Bowman, J.L., Eshed, Y., and Baum, S.F. (2002). Establishment of polarity in angiosperm lateral organs. *Trends Genet* 18 (3): 134–141.

Chen, J., Wang, S., Wu, F. et al. (2022). Genome-wide identification and functional characterization of auxin response factor (ARF) genes in eggplant. *Int J Mol Sci* 23: 6219.

Ckurshumova, W., Smirnova, T., Marcos, D. et al. (2014). Irrepressible MONOPTEROS/ARF5 promotes de novo shoot formation. *New Phytol* 204: 556–566.

de Jong, M., Wolters-Arts, M., Feron, R. et al. (2009). The Solanum lycopersicum auxin response factor 7 (SlARF7) regulates auxin signaling during tomato fruit set and development. *Plant J* 57: 160–170.

de Jong, M., Wolters-Arts, M., Schimmel, B.C.J. et al. (2015). Solanum lycopersicum AUXIN RESPONSE FACTOR 9 regulates cell division activity during early tomato fruit development. *J Exp Bot* 66: 3405–3416.

Du, L., Bao, C., Hu, T., et al. (2016). SmARF8, a transcription factor involved in parthenocarpy in eggplant. *Mol Genet Genomics* 291 (1): 93–105.

Ellis, C.M., Nagpal, P., Young, J.C. et al. (2005). AUXIN RESPONSE FACTOR1 and AUXIN RESPONSE FACTOR2 regulate senescence and floral organ abscission in Arabidopsis thaliana. *Development* 132: 4563–4574.

Faivre-Rampant, O., Cardle, L., Marshall, D. et al. (2004). Changes in gene expression during meristem activation processes in Solanum tuberosum with a focus on the regulation of an auxin response factor gene. *J Exp Bot* 55: 613–622.

Goetz, M., Hooper, L.C., Johnson, S.D. et al. (2007). Expression of aberrant forms of AUXIN RESPONSE FACTOR8 stimulates parthenocarpy in Arabidopsis and tomato. *Plant Physiol* 145: 351–366.

Goetz, M., Vivian-Smith, A., Johnson, S.D., and Koltunow, A.M. (2006). AUXIN RESPONSE FACTOR8 is a negative regulator of fruit initiation in Arabidopsis. *Plant Cell* 18: 1873–1886.

Guan, X., Xu, T., Gao, S. et al. (2014). Temporal and spatial distribution of auxin response factor genes during tomato flower abscission. *J Plant Growth Regul* 33: 317–327.

Guilfoyle, T.J. and Hagen, G. (2007). Auxin response factors. *Curr Opin Plant Biol* 10: 453–460.

Guilfoyle, T.J., Ulmasov, T., and Hagen, G. (1998). The ARF family of transcription factors and their role in plant hormone-responsive transcription. *Cell Mol Life Sci* 54: 619–627.

Hagen, G. (2015). Auxin signal transduction. In: *Essays in Biochemistry* (ed. T. Guilfoyle and G. Hagen), 1–12.

Hao, Y., Hu, G., Breitel, D. et al. (2015). Auxin response factor SlARF2 is an essential component of the regulatory mechanism controlling fruit ripening in tomato. *PLOS Genet* 11.

Hardtke, C.S. and Berleth, T. (1998). The arabidopsis gene MONOPTEROS encodes a transcription factor mediating embryo axis formation and vascular development. *Embo J* 17: 1405–1411.

Harper, R.M., Stowe-Evans, E.L., Luesse, D.R. et al. (2000). The NPH4 locus encodes the auxin response factor ARF7, a conditional regulator of differential growth in aerial Arabidopsis tissue. *Plant Cell* 12: 757–770.

Hendelman, A., Buxdorf, K., Stav, R. et al. (2012). Inhibition of lamina outgrowth following Solanum lycopersicum AUXIN RESPONSE FACTOR 10 (SlARF10) derepression. *Plant Mol Biol* 78: 561–576.

Hu, W., Zuo, J., Hou, X. et al. (2015). The auxin response factor gene family in banana: genome-wide identification and expression analyses during development, ripening, and abiotic stress. *Front Plant Sci* 6.

Ishibashi, N., Machida, C., and Machida, Y. (2013). ASYMMETRIC LEAVES2 and FASCIATA2 cooperatively regulate the formation of leaf adaxial-abaxial polarity in Arabidopsis thaliana. *Plant Biotechnol-Nar* 30: 411–415.

Iwasaki, M., Takahashi, H., Iwakawa, H. et al. (2013). Dual regulation of ETTIN (ARF3) gene expression by AS1-AS2, which maintains the DNA methylation level, is involved in stabilization of leaf adaxial-abaxial partitioning in Arabidopsis. *Development* 140: 1958–1969.

Jones, B., Frasse, P., Olmos, E. et al. (2002). Down-regulation of DR12, an auxin-response-factor homolog, in the tomato results in a pleiotropic phenotype including dark green and blotchy ripening fruit. *Plant J* 32: 603–613.

Kalluri, U.C., DiFazio, S.P., Brunner, A.M., and Tuskan, G.A. (2007). Genome-wide analysis of Aux/IAA and ARF gene families in Populus trichocarpa. *BMC Plant Biol* 7.

Klee, H.J. and Giovannoni, J.J. (2011). Genetics and control of tomato fruit ripening and quality attributes. In: *Annual Review of Genetics* (ed. B.L. Bassler, M. Lichten, and G. Schupbach), 41–59.

Kobayashi-Uehara, A., Shimosaka, E., and Handa, H. (2001). Cloning and expression analyses of cDNA encoding an ADP-ribosylation factor from wheat: tissue-specific expression of wheat ARF. *Plant Sci* 160: 535–542.

Kong, X., Zeng, J., Yun, Z. et al. (2022). Characterization of miRNA-mediated auxin signaling during banana (Musa spp.) fruit ripening. *Postharvest Biol Tec* 193: 112045.

Koornneef, M. and Meinke, D. (2010). The development of Arabidopsis as a model plant. *Plant J* 61: 909–921.

Korasick, D.A., Chatterjee, S., Tonelli, M. et al. (2015). Defining a two-pronged structural model for PB1 (Phox/Bem1p) domain interaction in plant auxin responses. *J Biol Chem* 290: 12868–12878.

Kumar, R., Tyagi, A.K., and Sharma, A.K. (2011). Genome-wide analysis of auxin response factor (ARF) gene family from tomato and analysis of their role in flower and fruit development. *Mol Genet Genomics* 285: 245–260.

Le, B., Nawaz, M.A., Rehman, H.M. et al. (2016). Genome-wide characterization and expression pattern of auxin response factor (ARF) gene family in soybean and common bean. *Genes Genom* 38: 1165–1178.

Li, G., Liang, W., Zhang, X. et al. (2014). Rice actin-binding protein RMD is a key link in the auxin-actin regulatory loop that controls cell growth. *P Natl Acad Sci USA* 111: 10377–10382.

Li, J.S., Dai, X.H., and Zhao, Y.D. (2006). A role for auxin response factor 19 in auxin and ethylene signaling in Arabidopsis. *Plant Physiol* 140: 899–908.

Li, S., OuYang, W., Hou, X. et al. (2015). Genome-wide identification, isolation and expression analysis of auxin response factor (ARF) gene family in sweet orange (Citrus sinensis). *Front Plant Sci* 6.

Li, X., Zhang, X., Shi, T. et al. (2022). Identification of ARF family in blueberry and its potential involvement of fruit development and pH stress response. *BMC Genomics* 23.

Liu, N., Wu, S., Van Houten, J. et al. (2014a). Down-regulation of AUXIN RESPONSE FACTORS 6 and 8 by microRNA 167 leads to floral development defects and female sterility in tomato. *J Exp Bot* 65: 2507–2520.

Liu, P., Montgomery, T.A., Fahlgren, N. et al. (2007). Repression of AUXIN RESPONSE FACTOR10 by microRNA160 is critical for seed germination and post-germination stages. *Plant J* 52: 133–146.

Liu, S., Zhang, Y., Feng, Q. et al. (2018). Tomato AUXIN RESPONSE FACTOR 5 regulates fruit set and development via the mediation of auxin and gibberellin signaling. *Sci Rep-UK* 8.

Liu, S.Q. and Hu, L.F. (2013). Genome-wide analysis of the auxin response factor gene family in cucumber. *Genet Mol Res* 12: 4317–4331.

Liu, X., Dinh, T.T., Li, D. et al. (2014b). AUXIN RESPONSE FACTOR 3 integrates the functions of AGAMOUS and APETALA2 in floral meristem determinacy. *Plant J* 80: 629–641.

Luo, X., Sun, M., Xu, R. et al. (2014). Genomewide identification and expression analysis of the ARF gene family in apple. *J Genet* 93: 785–797.

Mariotti, L., Picciarelli, P., Lombardi, L., and Ceccarelli, N. (2011). Fruit-set and early fruit growth in tomato are associated with increases in indoleacetic acid, cytokinin, and bioactive gibberellin contents. *J Plant Growth Regul* 30: 405–415.

Miao, H., Sun, P., Liu, Q. et al. (2020). Molecular identification of the key starch branching enzyme-encoding gene SBE2.3 and its interacting transcription factors in banana fruits. *Hortic Res-England* 7.

Nagpal, P., Ellis, C.M., Weber, H. et al. (2005). Auxin response factors ARF6 and ARF8 promote jasmonic acid production and flower maturation. *Development* 132: 4107–4118.

Okushima, Y., Fukaki, H., Onoda, M. et al. (2007). ARF7 and ARF19 regulate lateral root formation via direct activation of LBD/ASL genes in Arabidopsis. *Plant Cell* 19: 118–130.

Okushima, Y., Overvoorde, P.J., Arima, K. et al. (2005). Functional genomic analysis of the AUXIN RESPONSE FACTOR gene family members in Arabidopsis thaliana: unique and overlapping functions of ARF7 and ARF19. *Plant Cell* 17: 444–463.

Pekker, I., Alvarez, J.P., and Eshed, Y. (2005). Auxin response factors mediate Arabidopsis organ asymmetry via modulation of KANADI activity. *Plant Cell* 17: 2899–2910.

Qi, Y., Wang, S., Shen, C. et al. (2012). OsARF12, a transcription activator on auxin response gene, regulates root elongation and affects iron accumulation in rice (Oryza sativa). *New Phytol* 193: 109–120.

Qiao, Z., Ji, X., Li, H. et al. (2022). MdARF8: an auxin response factor involved in jasmonate signaling pathway in malus domestica. *J Plant Growth Regul.* doi: 10.1007/s00344-022-10656-8.

Qin, Q., Li, G., Jin, L. et al. (2020). Auxin response factors (ARFs) differentially regulate rice antiviral immune response against rice dwarf virus. *PLOS Pathog* 16: e1009118.

Ren, Z., Li, Z., Miao, Q. et al. (2011). The auxin receptor homologue in Solanum lycopersicum stimulates tomato fruit set and leaf morphogenesis. *J Exp Bot* 62: 2815–2826.

Rena, Z., Liu, R., Gu, W., and Dong, X. (2017). The Solanum lycopersicum auxin response factor SlARF2 participates in regulating lateral root formation and flower organ senescence. *Plant Sci* 256: 103–111.

Sato, Y., Nishimura, A., Ito, M. et al. (2001). Auxin response factor family in rice. *Genes Genet Syst* 76: 373–380.

Shen, C., Yue, R., Yang, Y. et al. (2014). OsARF16 is involved in cytokinin-mediated inhibition of phosphate transport and phosphate signaling in rice (Oryza sativa L.). *PLOS One* 9.

Si, F., Yang, C., Yan, B. et al. (2022). Control ofOsARF3a byOsKANADI1 contributes to lemma development in rice. *Plant J* 110: 1717–1730.

Song, X., Wang, D., Ma, L., et al. (2012). Rice RNA-dependent RNA polymerase 6 acts in small RNA biogenesis and spikelet development. *Plant J* 71 (3): 378–389.

Tang, Y., Bao, X., Liu, K. et al. (2018). Genome-wide identification and expression profiling of the auxin response factor (ARF) gene family in physic nut. *PLOS One* 13.

Tatematsu, K., Kumagai, S., Muto, H. et al. (2004). MASSUGU2 encodes Aux/IAA19, an auxin-regulated protein that functions together with the transcriptional activator NPH4/ARF7 to regulate differential growth responses of hypocotyl and formation of lateral roots in Arabidopsis thaliana. *Plant Cell* 16: 379–393.

Tian, C., Muto, H., Higuchi, K. et al. (2004). Disruption and overexpression of auxin response factor 8 gene of Arabidopsis affect hypocotyl elongation and root growth habit, indicating its possible involvement in auxin homeostasis in light condition. *Plant J* 40: 333–343.

Tiwari, S.B., Hagen, G., and Guilfoyle, T. (2003). The roles of auxin response factor domains in auxin-responsive transcription. *Plant Cell* 15: 533–543.

Tombuloglu, H. (2019). Genome-wide analysis of the auxin response factors (ARF) gene family in barley (Hordeum vulgare L.). *J Plant Biochem Biot* 28: 14–24.

Ulmasov, T., Hagen, G., and Guilfoyle, T.J. (1997). *ARF1, a Transcription Factor that Binds to Auxin Response Elements*, vol. 276, 1865–1868. New York, N.Y.: Science.

Ulmasov, T., Hagen, G., and Guilfoyle, T.J. (1999). Activation and repression of transcription by auxin-response factors. *P Natl Acad Sci USA* 96: 5844–5849.

Varaud, E., Brioudes, F., Szecsi, J. et al. (2011). AUXIN RESPONSE FACTOR8 regulates arabidopsis petal growth by interacting with the bHLH transcription factor BIGPETALp. *Plant Cell* 23: 973–983.

Waller, F., Furuya, M., and Nick, P. (2002). OsARF1, an auxin response factor from rice, is auxin-regulated and classifies as a primary auxin responsive gene. *Plant Mol Biol* 50: 415–425.

Wang, D., Pei, K., Fu, Y. et al. (2007). Genome-wide analysis of the auxin response factors (ARF) gene family in rice (Oryza sativa). *Gene* 394: 13–24.

Wang, J.W., Wang, L.J., Mao, Y.B. et al. (2005a). Control of root cap formation by microRNA-targeted auxin response factors in Arabidopsis. *Plant Cell* 17: 2204–2216.

Wang, S.C., Tiwari, S.B., Hagen, G., and Guilfoyle, T.J. (2005b). AUXIN RESPONSE FACTOR7 restores the expression of auxin-responsive genes in mutant Arabidopsis leaf mesophyll protoplasts. *Plant Cell* 17: 1979–1993.

Wang, S., Zhang, S., Sun, C. et al. (2014). Auxin response factor (OsARF12), a novel regulator for phosphate homeostasis in rice (Oryza sativa). *New Phytol* 201: 91–103.

Wang, Y., Deng, D., Shi, Y., et al. (2012). Diversification, phylogeny and evolution of auxin response factor (ARF) family: insights gained from analyzing maize ARF genes. *Mol Biol Rep* 39 (3): 2401–2415.

Wei, L., Xinyu, L., Ya, D. et al. (2022). Effects of exogenous ABA on the expression of resistance-related genes in tomato (Solanum lycopersicum). *Chin J Appl Environ Biol* 28: 582–587.

Wen, J., Guo, P., Ke, Y. et al. (2019). The auxin response factor gene family in allopolyploid Brassica napus. *PLOS One* 14.

Wenzel, C.L., Schuetz, M., Yu, Q., and Mattsson, J. (2007). Dynamics of MONOPTEROS and PIN-FORMED1 expression during leaf vein pattern formation in Arabidopsis thaliana. *Plant J* 49: 387–398.

Woodward, A.W. and Bartel, B. (2005). A receptor for auxin. *Plant Cell* 17: 2425–2429.

Wu, B., Wang, L., Pan, G. et al. (2020). Genome-wide characterization and expression analysis of the auxin response factor (ARF) gene family during melon (Cucumis melo L.) fruit development. *Protoplasma* 257: 979–992.

Wu, J., Wang, F., Cheng, L. et al. (2011). Identification, isolation and expression analysis of auxin response factor (ARF) genes in Solanum lycopersicum. *Plant Cell Rep* 30: 2059–2073.

Wu, M., Tian, Q., and Reed, J.W. (2006). Arabidopsis microRNA167 controls patterns of ARF6 and ARF8 expression, and regulates both female and male reproduction. *Development* 133: 4211–4218.

Xing, H., Pudake, R.N., Guo, G. et al. (2011). Genome-wide identification and expression profiling of auxin response factor (ARF) gene family in maize. *BMC Genomics* 12.

Xu, Q., Chen, L., Ruan, X. et al. (2013). The draft genome of sweet orange (Citrus sinensis). *Nat Genet* 45: 59–92.

Xu, X., Zhang, Q., Gao, X. et al. (2022). Auxin and abscisic acid antagonistically regulate ascorbic acid production via the SlMAPK8–SlARF4–SlMYB11 module in tomato. *Plant Cell* 34: 4409–4427.

Xue, H., Meng, J., Lei, P. et al. (2022). ARF2-PIF5 interaction controls transcriptional reprogramming in the ABS3-mediated plant senescence pathway. *EMBO J* 41 (19): e110988.

Yamasaki, K., Kigawa, T., Inoue, M., et al. (2004). A novel zinc-binding motif revealed by solution structures of DNA-binding domains of Arabidopsis SBP-family transcription factors. *J Mol Biol* 337 (1): 49–63.

Yang, J., Tian, L., Sun, M. et al. (2013). AUXIN RESPONSE FACTOR17 Is Essential for Pollen Wall Pattern Formation in Arabidopsis. *Plant Physiol* 162: 720–731.

Yifhar, T., Pekker, I., Peled, D. et al. (2012). Failure of the tomato trans-acting short interfering RNA program to regulate AUXIN RESPONSE FACTOR3 and ARF4 underlies the wiry leaf syndrome. *Plant Cell* 24: 3575–3589.

Yoon, E.K., Yang, J.H., and Lee, W.S. (2010). Auxin and abscisic acid responses of auxin response factor 3 in arabidopsis lateral root development. *J Plant Biol* 53: 150–154.

Yuan, Y., Mei, L., Wu, M. et al. (2018). SlARF10, an auxin response factor, is involved in chlorophyll and sugar accumulation during tomato fruit development. *J Exp Bot* 69 (22): 5507–5518.

Yuan, Y., Xu, X., Gong, Z. et al. (2019). Auxin response factor 6A regulates photosynthesis, sugar accumulation, and fruit development in tomato. *Hortic Res-England* 6.

Zhang, H., Cao, N., Dong, C., and Shang, Q. (2017). Genome-wide identification and expression of ARF gene family during adventitious root development in hot pepper (Capsicum annuum). *Horticult Plant J* 3: 151–164.

Zhang, L., Guan, R., Li, G. et al. (2022). Genomic identification of ARF transcription factors and expression analysis in Cannabis sativa L. *Ind Crop Prod* 186: 115118.

Zhang, S., Wang, S., Xu, Y. et al. (2015a). The auxin response factor, OsARF19, controls rice leaf angles through positively regulating OsGH3-5 and OsBRI1. *Plant Cell Environ* 38: 638–654.

Zhang, X., Yan, F., Tang, Y. et al. (2015b). Auxin response gene SlARF3 plays multiple roles in tomato development and is involved in the formation of epidermal cells and trichomes. *Plant Cell Physiol* 56: 2110–2124.

Zhang, S., Wu, T., Liu, S. et al. (2016). Disruption of OsARF19 is critical for floral organ development and plant architecture in rice (Oryza sativa L.). *Plant Mol Biol Rep* 34: 748–760.

Zheng, L., Zhou, J., Yi, L., and Zhou, X. (2021). Identification and expression analysis of the auxin response factor (ARF) gene family in sorghum [Sorghum bicolor (L.) Moench]. *Pak J Bot* 53: 905–914.

Zouine, M., Fu, Y., Chateigner-Boutin, A. et al. (2014). Characterization of the tomato ARF gene family uncovers a multi-levels post-transcriptional regulation including alternative splicing. *PLOS One* 9.

7

HD-ZIPs Are Involved in Flesh Fruit Development and Ripening

7.1 Introduction

Homeodomain-leucine zipper (HD-Zip) proteins are a plant-specific transcription factor family (Elhiti and Stasolla 2009) widely present in most plants from mosses to higher plants, including *Arabidopsis thaliana* (Sorin et al. 2009; Son et al. 2010; Peterson et al. 2013), rice (*Oryza sativa* L.), maize (*Zea Mays* L.) (Jiao et al. 2022; Qiu et al. 2022), tomato (*Solanum lycopersicum*) (Liu et al. 2022), and poplar (*Populus trichocarpa T&G*). Using whole-genome sequence analysis, 47 HD-Zip transcription factors have been identified in *Arabidopsis*, 33 in rice, 55 in maize, 58 in tomato, and 63 in poplar. Bayarsaihan et al. (2003) reported an atypical vertebrate HD-Zip transcription factor, designated HD-Zip-encoding gene (*Homez*); however, sequence alignment revealed that the leucine zipper (LZ) motifs in Homez are remarkably different from typical LZ motifs in HD-Zip proteins and other plant LZ-containing proteins.

7.2 Structure of HD-Zip Genes

The proteins encoded by HD-Zip genes contain a C-terminal homeodomain (HD) consisting of 60 (or 61) highly conserved amino acid residues. The protein motif folds into three-stranded α-helices separated by a loop and a turn, forming a typical helix-loop-helix-turn-helix structure, and its third α-helix lies within the major groove of DNA (Chan et al. 1998). The third α-helix, the variable N-terminal arm, and the loop between the first two α-helices are the main structures responsible for the specific recognition and binding of HD-Zip transcription factors to their target genes; in particular, these structures are closely associated with the specific recognition of the cis-acting elements in the 5′ upstream region of the target genes (Elhiti and Stasolla, 2009).

Transcriptional Regulation of Flesh Fruit Development and Ripening, First Edition. Yunbo Luo.

An LZ domain is located immediately C-terminal to the homeodomain. This LZ structure is characterized by the presence of a leucine (Leu) residue or methionine (Thr) residue every seven amino acids. The leucine residue sidechains protrude from and are regularly arranged on one side of the polypeptide chain that forms an α-helix, and two α-helices are interleaved through leucine residues to form a stable zipper-like hydrophobic domain. Proteins containing such a structure often bind to DNA sequences by forming homo- or heterodimers and would lose their binding ability were they lacking the LZ structure (Shaikhali et al. 2012).

7.3 HD-Zip Gene Classification

All HD-Zip family genes have an HD domain and an LZ motif (Zip) (Zhu et al. 2022). They can be divided into four subclasses, Class I, II, III, and IV (Wang et al. 2022b), based on the degrees of homology and conservation of the HD-Zip sequence, domain similarity, and physiological function.

Most HD-Zip Class I proteins consist of approximately 300 amino acids with a molecular weight of approximately 35 kDa and an HD-LZ domain in the middle. Except for the highly conserved HD-Zip domain, other regions in HD-Zip I proteins vary in the degree of conservation. Among the family members, the LZ motif shows remarkable sequence differences, while the N-terminal and C-terminal sequences have low degrees of homology, which suggests that different HD-Zip I family members may interact with different proteins.

In *A. thaliana*, the Class I HD-Zip transcription factor family has 17 genes. Using in vitro polymerase chain reaction (PCR)-assisted binding site selection and footprinting assays, researchers confirmed that the proteins encoded by the HD-Zip Class I genes can form dimers and recognize the complementary palindromic sequence CAAT(A/T) ATTG (Chan et al. 1998; Palena et al. 1999, 2001), which is named homeodomain recognition element HDE1. Among the HD-Zip Class I members, only AtHB7 and AtHB12 cannot bind to DNA in vitro (Johannesson et al. 2001). In vitro experiments on sunflower (*Helianthus annuus*) HD-Zip Class I proteins showed that the variable N-terminal arm plays a significant role in the binding of HD-Zip proteins with DNA (Palena et al. 2001) and can increase the binding affinity without affecting the DNA-binding specificity. In addition, the inability of HD-Zip proteins to bind DNA in their monomeric form as the majority of HD proteins is related to the composition of the loop located between the first and second helices.

HD-Zip II genes are similar in size to that of Class I genes. In addition to the highly conserved HD and LZ motif, HD-Zip II proteins have two conserved Cys-Pro-Ser-Cys-Glu (CPSCE) motifs that are located downstream of the LZ motif (i.e., the C-terminus) and involved in cellular redox potentials. In in vitro experiments, HD-Zip II transcription factors formed high-molecular-weight polymers through intermolecular cysteine–cysteine crosslinking under oxidation conditions, and these polymers cannot be transported into the nuclei to function. This finding suggests that cellular redox status may determine the activity of HD-Zip II proteins (Tron et al. 2002).

In *A. thaliana*, the HD-Zip II transcription factor family has 17 members. These proteins bind specifically to the complementary palindromic sequence CAAT(G/C) ATTG (HDE2 or AH2), for which only two bases at the central position of the recognition sequence differ from those in the DNA sequence specifically bound by HD-Zip I proteins. The specificity for binding at the central position is determined by amino acids 46 and 56 of helix III. These two amino acids, along with the conserved Arg55, determine the proteins with different spatially symmetrical structures, and Arg55 is directly responsible for the interaction (Tron et al. 2005).

HD-Zip III proteins differ greatly from HD-Zip I and II in structure. Similar to the other two classes, HD-Zip III proteins have an HD-LZ domain, but they have at least four additional amino acids inserted in both the HD and LZ motifs, a steroidogenic acute regulatory protein-related lipid transfer (START) domain, and a START-adjacent domain (SAD). The START domain is phylogenetically highly conserved, suggesting that it might play an important regulatory role. This domain contains approximately 200 amino acids and can bind hydrophobic small molecules, such as steroids, phospholipids, and carotenoids. HD-Zip Class III members have a conserved methionine-glutamate-lysine-histidine-leucine-alanine (MEKHLA) domain at the C-terminus (Yang et al. 2022). This domain is similar to the Per-ARNT-Sim (PAS) domain, which can be found in many proteins throughout all kingdoms of life and participates in light, oxygen, and redox potential sensation and the reception and transmission of external signals (Mukherjee et al. 2006; Möglich et al. 2009; McIntosh et al. 2010). Five HD-Zip III transcription factors have been found in *A. thaliana*. Although only a few studies have investigated the DNA sequence bound to HD-Zip III proteins in *Arabidopsis*, ATHB9 protein was found to have a high affinity for the GTAAT(G/C) ATTAC sequence in vitro.

Similar to HD-Zip III proteins, HD-Zip IV proteins have HD, LZ, START, and SAD domains (Ahmad et al. 2022); however, they do not have a MEKHLA domain. These proteins are distinctly characterized by a loop structure in the middle of the LZ motif. After the discovery of GLABRA2 protein in *A. thaliana*, HD-Zip IV proteins were named HD-Zip GL2 (GL2 family) (Nakamura et al. 2006). These transcription factors have similar target binding sequences and dimeric structures to those of HD-Zip I and II proteins and can specifically bind to the CATT(A/T) AATG sequence. Class IV proteins appear to bind alternative sequences more easily. In sunflower, CATT(A/T)AATG is the preferred binding sequence for HD-Zip IV protein HAHR1 (Tron et al. 2001). The binding site selection experiment revealed that HDG7, HDG9, ATML1, and PDF2a have a common binding sequence GCATT(A/T)AATGC. In vitro experiments on ATML1 further showed that this binding sequence is overlapping with the L1 box sequence TAAATG(C/T)A (Abe et al. 2001). In addition, the available data demonstrated that the target sequence of HD-Zip IV proteins is characterized by a TAAA core. This characteristic structure exists in the Dj1 gene promoter and is the target of GL2 (Ohashi and Fukuda 2003).

7.4 Mechanism of Action of HD-Zip Transcription Factors

The featured domain structure is the foundation for HD-Zip transcription factors to interact with other proteins. Two members of *Arabidopsis* HD-Zip Class I, AtHB7 and AtHB12, regulate the abscisic acid (ABA) signaling pathway by acting on protein phosphatase 2C (PP2C) and repressing the activity of ABA receptors (Valdés et al. 2012). AtHB6 is a negative regulator of the ABA signal pathway. Its overexpression decreases ABA sensitivity and suppresses the expression of RAR18, RD22, and RD29B in plants. PYR/RCAR, together with PP2C and SnRK2s, regulates the ABA signaling pathway. AtHB6 also interacts with PP2C phosphorylase ABI, indicating that AtBH6 is located upstream of the ABA signal (Himmelbach et al. 2002). MATH-BTB protein is a substrate-binding converter in the ABA signaling pathway and can directly bind with E3 ubiquitin ligase to form CUL3 BPM complexes. Studies have shown that MATH-BTB interacts with AtHB6 to degrade AtHB6 protein. A decrease in CUL3 BPM complexes can lead to AtHB6 protein accumulation, suppress plant growth and pollination, and affect the response to ABA and stomatal behavior of plants (Lechner et al. 2011). In addition, AtHB6 binds to *cis*-elements to activate transcription of its own encoding gene or other genes (Soderman et al. 1999). However, exogenous ABA could eliminate the effect of AtHB6 overexpression on AtHB6 gene expression. This effect suggests that AtHB6 itself is insufficient to trigger transcriptional activity, which might depend on other transcriptional mechanisms, e.g., it activates other genes, especially other HD-Zip transcription factors, to inhibit the expression of downstream genes (Lechner et al. 2011).

7.5 Functions of the HD-Zip Family

7.5.1 HD-Zip Class I

All HD-Zip I members have similar developmental relationships and similar expression patterns in the same organs and respond to the same environmental factors (Henriksson et al. 2005). During the developmental process, these transcription factors respond to various stimuli, such as drought, high salt, low temperature, high temperature, light changes, fungal and bacterial infections, and other environmental stresses, especially certain abiotic factors, to improve the adaptability of plants to environmental stresses. Some adaptive responses occur specifically in certain tissues and organs of plants.

1) Involvement in environmental stress response

When plants face environmental stress, some transcription factors regulate endogenous gene expression by sensing exogenous environmental signals. ABA is an important hormone that participates in plant growth and development and regulates plant response to biotic and abiotic stresses, especially drought, salt stress, and low temperature (Deng et al. 2006).

HD-Zip Class I members may be involved in ABA regulation and abiotic stress response (Li et al. 2022). Studies on gene expression and transgenic plants have found that AtHB7 and AtHB12 are upregulated and AtHB5 and AtHB6 are downregulated under the conditions of water shortage and/or ABA treatment. In response to these stimuli, HD-Zip I genes demonstrate a regulatory effect on plant development and growth.

AtHB12 is an important active transcriptional factor that participates in not only the regulation of specific development processes but also the ABA-regulated gene expression response to water stress in plants. Lee et al. (2001) fused the 5′-terminal regulatory region of *AtHB12* with the GUS reporter gene and then detected the expression pattern of GUS in transgenic plants. They found that, under normal growth conditions, the *AtHB12*::GUS fusion gene was weakly expressed in the apical meristem region. In contrast, ABA treatment strongly induced GUS expression in the stem, petiole, leaves around the apical meristem, and elongation regions of roots in transgenic tobacco with the *AtHB12*::GUS fusion gene. These results indicate that there are ABA *cis*-acting elements in the *AtHB12* promoter region. Moreover, the expression pattern of *AtHB12* is complex and often associated with regions of developing vascular tissues.

AtHB6 is constitutively expressed in seedlings, and its expression is significantly upregulated under water-limited conditions, osmotic stress, or exogenous ABA treatment. The transcription level of *AtHB6* increased threefold in drought or high-salt conditions and increased fivefold after 10 μM ABA treatment, consistent with the ABA-mediated responses to water stress and osmotic stress (Söderman et al. 1999).

AtHB7 transcripts appear at a low level in all plant organs. Water-limited stress, osmotic stress, and exogenous ABA treatments can increase AtHB7 expression by several-fold. The response began to be detectable when the plant was treated with exogenous 10^{-8} M ABA and reached the maximum when treated with 10^{-6} M ABA. *AtHB7* transcripts could be detected after 30 minutes of ABA treatment and decreased rapidly after the removal of ABA exposure. *AtHB7* expression can be directly induced by ABA; in the ABA-deficient mutant *aba-3*, drought stress treatment failed to induce *AtHB7* expression. In the *aba-1* mutant, the ABA concentration required to maximally induce AtHB7 expression was 100 times that for wild-type plants. The *aba-1* mutant plants showed weaker AtHB7 response to drought stress and osmotic stress treatment. These results suggest that ATHB-7 is transcriptionally regulated in an ABA-dependent manner and may act in a signal transduction pathway mediating a drought response and including ABI1 (Söderman et al. 1996).

AtHB5 belongs to *Arabidopsis* HD-Zip Class I. In experiments where plants were treated with different concentrations of ABA, the seed germination rate was significantly lower in transgenic *Arabidopsis* plants transfected with a *35S::ATHB5* construct than that of wild-type plants. In experiments analyzing the dose-response pattern of ABA sensitivity of root growth, the *35S::AtHB5* seedlings showed higher sensitivity to ABA at all concentrations when compared with wild-type seedlings. Research has shown that enhancing *AtHB5* gene expression could improve ABA sensitivity in transgenic *Arabidopsis* plants. *AtHB5* positively affects plant response to ABA and thus regulates the inhibitory effect of ABA during seedling growth (Henrik et al. 2003).

Nicotiana attenuata NaHD20 belongs to HD-Zip Class I, and its expression is induced by stress factors such as drought and injury. In a phytohormone analysis study where *NaHD20* gene expression in *N. attenuata* leaves was silenced by using virus-induced gene silencing (VIGS) technology, the leaf ABA content increased under water stress and reached the peak after 18 days of treatment. The water stress-induced ABA increase in leaves was less prominent in the control group than that in the experimental group (the *NaHD20*-silenced group). Recovery from water deficiency resulted in a significant decrease in the ABA content and upregulation of some dehydration-responsive genes, including ABA biosynthetic genes. Therefore, *NaHD20* plays a promotive role in ABA accumulation in leaves of water-stressed plants (Ré et al. 2011).

OsHOX6, *OsHOX22*, and *OsHOX24* in rice (*Oryza sativa*, *Os*) are orthologous to *Arabidopsis AtHB7* and *AtHB12*, and their expression levels are upregulated under drought conditions. These three genes vary in the expression pattern in rice and differ from their orthologous genes in *Arabidopsis thaliana*. Phylogenetic analysis confirmed that rice *Oshox22* and *Arabidopsis AtHB7* and *AtHB12* are HD-Zip I genes, and their expression could be induced by drought, salt, ABA, and polyethylene glycol (PEG) (Agalou et al. 2008). The *Populus* transcription factors homologous to *Arabidopsis AtHB7* play an important role in maintaining cellular osmotic pressure under salt stress conditions. Sunflower *HaHB4* is also a drought-induced ABA-responsive gene that has a sequence highly similar to that of *Arabidopsis AtHB7* and *AtHB12*; its expression is induced by high salt, water deficit, and ABA (Dezar et al. 2005). *HaHB4* promoter contains an ABA-response element and a NaCl-response element. *Arabidopsis* plants overexpressing *HaHB4* demonstrated the phenotypes of strong drought tolerance, high-salt tolerance, and delayed senescence (Manavella et al. 2006, 2008).

Plants respond to cold stress through different signal transduction pathways. Two HD-Zip Class I transcription factors, sunflower *HaHB1* and *Arabidopsis AtHB13*, play an important role in cold stress response. The expression levels of these two genes are upregulated under low-temperature conditions. When these two genes were constitutively expressed in *Arabidopsis*, the transgenic plants showed similar phenotypes, including cell membrane stabilization under freezing treatments and cold stress conditions. A transcriptomic analysis of *HaHB1* transgenic plants indicated that various transcripts encoding glucanases and chitinases were induced. Moreover, under freezing conditions, some proteins accumulated in apoplasts of *HaHB1* plants, and these extracts exerted antifreeze activity in vitro. The *Arabidopsis* transgenic plants overexpressing two glucanase-encoding genes and one chitinase-encoding gene exhibited cell membrane stabilization after a short freezing treatment. Therefore, *HaHB1* and *AtHB13* are involved in plant cold tolerance via the induction of proteins able to stabilize cell membranes and inhibit ice growth (Cabello et al. 2012).

Under environmental stress, plants can adapt to and grow in the adverse environment by changing the architecture of the root system and promoting lateral root formation. Salt stress induces the expression of *MtHB1* gene, an HD-Zip I transcription factor, in main and lateral root tissues in *Medicago truncatula*. *Medicago* plants overexpressing *MtHB1* gene showed changes in the architecture of the root system with a larger number of lateral roots. HB1 recognizes the *cis*-element CAATAATTG present in the lateral organ boundaries (LOB)-binding domain (*LBD*) promoter region of LOB-like transcription factors and inhibits *LBD* gene expression. Hence, it is speculated that *MtHB1* may reduce root surface exposure to an adverse environment by regulating root extension and inhibiting lateral root growth, which improves the adaptability of the plants to the adverse environment. Auxin, a key hormone involved in lateral root formation, regulates pericycle cells by controlling their asymmetric transverse division, primordial boundary formation, and root formation. ABA regulates lateral root differentiation and formation by acting on ABI3 protein in *Arabidopsis* (Brady et al. 2003). The findings above show that HD-Zip transcription factors improve plant resistance to abiotic stress through the cross-interaction of auxin and ABA.

HaHB4 is a sunflower HD-Zip I gene and is transcriptionally regulated by water availability and ABA. *HaHB4* participates in the novel mechanism of ethylene-regulated senescence and thus plays a role in improving dehydration tolerance of plants. Transgenic *Arabidopsis* plants expressing this gene under the control of its own inducible promoter exhibited high water stress tolerance and an inverse correlation between ethylene sensitivity and *Hahb4* levels. *Arabidopsis* plants overexpressing Hahb-4 were less sensitive to external ethylene, entered the senescence pathway later, and did not show the typical triple response. Furthermore, *HaHB4* overexpression showed a significant repressive effect on genes related to ethylene synthesis and signaling, including *ACO*, *SAM*, *ERF2*, and *ERF5* (Manavella et al. 2008).

Gibberellins (GAs) can induce the upregulation of *HaHB10* expression. In transgenic *Arabidopsis* plants overexpressing *HaHB10*, the leaf shape, color, plant growth rate, flowering time, and life cycle were changed, and these phonotypic changes could be reversed by exogenous GA treatment. These results indicate that *HaHB10* is involved in the GA-mediated signal transduction pathway. HaHB10 regulates plant responses to biotic stresses through the pathway involving both salicylic acid (SA) and jasmonic acid (JA) (Rueda et al. 2005; Dezar et al. 2011).

The combination of the HDZI promoter with DREB/CBF genes can be used in transgenic cereals to improve tolerance to abiotic stresses and to reduce the negative effects of transgenes on cereal growth and production (Yang et al. 2020).

2) Involvement in photomorphogenesis

In nature, light stimulates seedlings to de-etiolate. In the late stage of seed germination, a far-red-rich light source can stimulate the phytochrome signaling system and induce a shade avoidance response (Stamm and Kumar 2010). The expression levels of AtHB3 and AtHB23 increased under dark conditions, while AtHB20 expression increased instantaneously in seeds after growing under light for three hours. Tobacco seedlings overexpressing *AtHB1* de-etiolated and could not form palisade cells when cultured in darkness (Aoyama et al. 1995). *AtHB16* negatively regulates cotyledon extension. In transgenic plants expressing an *AtHB16* cDNA or an antisense *AtHB16* cDNA grown in blue light, the transcription level of *AtHB16* was negatively correlated with cotyledon length (Wang et al. 2003). Blue light and two cryptochrome genes, *CRY1* and *CRY2*, do not affect the expression of *AtHB1*, and *AtHB16* does not affect the expression of *CRY1* and *CRY2*. In plants grown in white light, *AtHB16* is highly expressed in leaves and negatively regulates leaf cell expansion, while expression of antisense *AtHB16* cDNA increases the leaf area (Wang et al. 2003). Blue light and darkness can also induce an upregulated expression of *AtHB52* (Henriksson et al. 2005). *AtHB20* integrates exogenous light signals and endogenous ABA signals to stimulate seed germination in *Arabidopsis* (Barrero et al. 2010). *ATHB20*, an HD-Zip I transcription factor, is expressed in the micropylar endosperm and the root cap of germinating seeds and participates in seed germination and ABA signal transduction. Transgenic *Arabidopsis AtHB20* mutant seeds have higher ABA sensitivity and increased seed dormancy (Barrero et al. 2010).

3) Participation in plant growth regulation

HD-Zip I proteins are involved in the regulation of fruit ripening (An et al. 2022). The tomato (*L. esculentum*) *LeHB1* gene is an HD-Zip I gene homologous to the *Arabidopsis AtHB1* gene. LeHB1 could recognize and bind to a 9–10 bp DNA sequence in the *LeACO1* promoter region to regulate *LeACO1* expression in vitro. In plants with VIGS-silenced *LeHB1* expression, *LeACO1* expression significantly decreased, leading to inhibition of fruit ripening. Overexpression of *LeHB-1* by viral delivery to developing flowers resulted in altered floral organ morphology, including production of multiple flowers within one sepal whorl, fusion of sepals and petals, and conversion of sepals into carpel-like structures that grew into fruits and ripened (Lin et al. 2008).

7.5.2 HD-Zip Class II

HD-Zip II proteins are involved in the regulation of flowering time. HaHB10 is a sunflower HD-Zip II transcription factor. Research has shown that HaHB10 induces flowering in plants. *Arabidopsis* plants overexpressing *HaHB10* showed a phenotype of early flowering and upregulated expression of the flowering-related genes *FT*, *FUL*, and *SEP3*. During the transition from the reproductive to flowering stage, *HaHB10* promotes early formation of flowers by regulating the expression of the abovementioned specific flowering transition genes. *HaHB10* not only regulates the transition from the vegetative to reproductive growth stage by inducing the expression of flowering-related genes but also regulates the hormone-mediated response to biological stress in plants (Dezar et al. 2011).

Five *Arabidopsis* HD-Zip Class II members are involved in photomorphogenesis (Ciarbelli et al. 2008). *AtHB2* and *HAT4* are highly expressed in etiolated seedlings grown in the darkness, and their expression can be strongly inhibited by red (R) and far-red (FR) light (Carabelli et al. 1996). Transgenic plants overexpressing *AtHB2/HAT4* exhibited longer hypocotyls and smaller cotyledons, while transgenic plants expressing an *AtHB2/HAT4* antisense construct had shorter hypocotyls (Rueda et al. 2005). *AtHB2/HAT4* expression could be enhanced by white light exposure and induced by a light source with a low R:FR ratio, leading to a plant shade avoidance response characterized by stem elongation and leaf reduction. Transgenic *AtHB2/HAT4* plants grown under white light showed decreased leaf size and inhibition of cell differentiation (Steindler et al. 1997). Ectopic overexpression of *HAT1* and *HAT2* derived similar phenotypes to those of *AtHB2/HAT4* transgenic plants. A light source rich in far-red also strongly induces the expression of *HAT1* and *HAT2* (Ciarbelli et al. 2008). Sunflower *HaHB10* is highly homologous in sequence to *AtHB2* and *HAT4*. Darkness strongly induces the upregulated expression of *HaHB10*, which is mainly expressed in mature photosynthetic organs (Rueda et al. 2005). Far-red light also induces the expression of *AtHB4* and *HAT3* (Ciarbelli et al. 2008). *AtHB4*-overexpressing transgenic plants had a phenotype similar to that of *AtHB2/HAT4* transgenic plants under white light conditions, but their far-red-induced hypocotyl elongation was suppressed (Sorin et al. 2009).

7.5.3 HD-Zip Class III

HD-Zip Class III members mainly regulate the development of the apical meristem, vascular bundles, and adaxial domains of lateral organs (Ariel et al. 2007).

1) Regulation of apical meristem formation

Plant growth essentially depends on cell division and differentiation of the apical meristem. Apical meristem includes shoot apical meristem (SAM) and root apical meristem (RAM), and SAM can differentiate into leaf primordia and floral primordia. In *Arabidopsis* and rice, except for *AtHB8* and *OsHB5*, all HD-Zip III genes are expressed in apical meristem, leaf primordia, floral primordia, and lateral root primordia (Williams et al. 2005). *Arabidopsis revoluta* (*rev*) mutants lack meristematic tissue in the leaf axils and have a significantly reduced number of paraclades, and their flowers often fail to complete normal development and form incomplete or abortive structures, suggesting that the *REV* gene plays an important role in SAM formation (Talbert et al. 1995; Otsuga et al. 2001). Both *PHABULOSA* (*PHB*) and *PHAVOLUTA* (*PHV*) can further enhance the phenotype of *REV* single mutant plants, suggesting that the three genes function synergistically in the formation of apical meristem. Triple *CAN/AtHB8/REV* mutants have larger numbers of paraclades and fertile flowers than *REV* single mutant plants, indicating that *CNA*, *AtHB8*, and *REV* have antagonistic roles in SAM formation (Prigge et al. 2005). Triple *PHB/PHV/REV*-deficient mutants have fewer lateral roots and the dominant mutants *rev-10d* have more lateral roots during RAM development, suggesting that *REV*, *PHB*, and *PHV* jointly promote RAM formation and function synergistically in the development of roots (Hawker and Bowman 2004).

2) Regulation of leaf polarity establishment

HD-Zip III genes are strongly expressed on the adaxial side of leaf primordia, suggesting that they play a role in the establishment of adaxial–abaxial polarity of leaves at leaf initiation. The *Arabidopsis* *REV*-dominant mutant *rev-10d/avb1* and *PHB*-dominant mutant *phb-1d* demonstrate very distinct leaf morphological traits. Their leaves have varying degrees of adaxial epidermal characters, which leads to rod-like, trumpet-like, or even needle-like leaf shapes. Observation of sectioned leaves revealed the adaxialization of the leaf vascular structure where xylem surrounded phloem tissue (McConnell and Barton 1998; Zhong and Ye 2004). The findings further confirm that this class of genes is essential for leaf polarization. The *CNA*-dominant mutant *icu4-1* demonstrates a less obvious polarity defect as that of *PHB*- and *REV*-dominant mutants, as evidenced by their upcurling leaves that have no significant vascular changes (Ochando et al. 2006). The *Arabidopsis* *REV*-deficient mutant *rev-1* has large, downcurling leaves. Anatomical observation showed that these leaves had a significantly larger number of mesophyll cells and cells in the vascular bundle; in particular, the adaxial face of leaves showed a more significant increase in the number and volume of cells (Talbert et al. 1995). The double gene-deficient mutant *rev/phv* has trumpet-like leaves, and their adaxial face displays more distinct abaxial epidermal characters. The leaves of the triple gene-deficient mutant *rev phb/+phv* almost all had a trumpet-like shape and were more seriously abaxialized (Prigge et al. 2005). As evidenced by the fact that the abovementioned dominant mutants and gene-deficient mutants have opposite phonotypes, HD-Zip Class III genes function as regulators in the establishment of leaf polarity. However, the mechanism underlying the effect of HD-Zip Class III genes on leaf polarization remains to be clarified.

The HD-ZIP III family transcription factor LF1 (lateral floret1) (OsHB1/HOX10) maintains the tissue development balance of the paraxial–paraxial axis in rice leaves by directly regulating *OsYUCCA6* expression and affecting the auxin distribution. Further studies found that LF1 can directly activate the expression of LITTLE ZIPPER family gene *OsZPR4* and HD-ZIP II family gene *OsHOX1*, and that OsZPR4 and OsHOX1 can interact with LF1 protein, respectively, to form heterodimers to inhibit the transcriptional activity of LF1 (Zhang et al. 2021).

3) Regulation of cell differentiation in the vascular system

As the main characteristic structure in terrestrial plants, the vascular system provides mechanical support for the upward growth of plants and transports water, nutrients, and some signaling molecules. Vascular bundles are composed of three orderly arranged tissues: xylem, procambium/cambium, and phloem.

In *Arabidopsis*, rice, and *Populus*, Class III HD-Zip genes are highly expressed in the vascular bundles of roots, stems, and leaves; in particular, five *Arabidopsis* HD-Zip III genes are specifically expressed in procambium and xylem cells (Zhong and Ye 1999). In the dominant mutants *rev-10d/avb1* and *phb-1d*, the vascular tissue changed dramatically, including the surrounding of phloem by xylem and a significant increase in the number of vascular bundles (McConnell et al. 2001; Kang and Dengler 2002; Zhong and Ye 2004). *REV* deficiency resulted in abnormal stem vascular bundles, loss of interfascicular fibers, and lodging tendency in *interfascicular fiberless1 (ifl1)* mutant ecotype Col (Columbia) plants (Zhong and Ye 1999). In *rev-1* mutants established from Ler (Landsberg *Erecta*) ecotype plants, the number of cells increased in the stem cortical layer, xylem, and phloem, and the cellular area in the cambium expanded (Talbert et al. 1995). The *CNA*-dominant mutant *icu4-1* had an increased number of cells in the xylem, cambium, and phloem (Ochando et al. 2006), indicating that *CNA* plays an opposite role to *REV* in the development of the vascular system. Baima et al. (2001) found that overexpressing *AtHB8* resulted in an increase in only the number of xylem cells in *Arabidopsis*, suggesting that *AtHB8* may function differently from several other members and specifically regulate xylem development (Baima et al. 2001). Compared to the single *REV*-deficient mutant *rev*, both double gene-deficient mutants *rev/phb* and *rev/phv* demonstrated more serious developmental defects of the vascular system, and the triple mutant *rev/cna/AtHB8* showed partially inhibited vascular developmental defects (Prigge et al. 2005). These findings suggest that *REV*, *PHB* and *PHV*, along with *CNA* and *AtHB8*, play antagonistic roles in stem vascular development in *Arabidopsis*. In *Arabidopsis* roots, the five HD-Zip III genes above are expressed in all types of vascular cells, including procambium cells, xylem precursor cells, protoxylem cells, and metaxylem cells, except for phloem cells. The metaxylem cells are replaced by ectopic protoxylem cells in *AtHB8/phb* and all triple gene-deficient mutants; metaxylem cells nearly completely disappear in quadruple gene-deficient mutants; and no xylem cells can be seen in mutants deficient in all five genes. These findings indicate that HD-Zip III genes determine the production of xylem, and their expression levels determine the formation of different types of xylem cells (Carlsbecker et al. 2010). Vascular development of woody plants differs from that of herbaceous plants. In woody plants, vascular cambium that develops on the base of the primary vascular system maintains a stem cell activity for a long time, allowing cells to continuously differentiate outward to form secondary phloem tissue and inward to form secondary xylem tissue. Research has confirmed the pivotal role of *PtHB7* in the differentiation and regulation of secondary xylem in poplar. In addition, *PtHB2* and *PtHB6*, which are homologous to *REV* and *CNA*, respectively, affect the development of primary and secondary xylem in poplar. *PtHB2* overexpression results in the presence of ectopic cambium and xylem cells, suggesting a regulatory role of *PtHB2* in cambium initiation. Overexpression of *PtHB6* inhibits the lignification in cells of secondary xylem and phloem fibers, suggesting that *PtHB6* plays a negative regulatory role in the development of secondary vascular tissue (Du et al. 2011; Robischon et al. 2011).

4) Involvement in the regulation of sepal development

HD-Zip III proteins also participate in the regulation of sepal development. Some studies have predicted the molecular regulatory network and proposed a developmental model of *Arabidopsis* sepals by using mathematical and computing methods (La Rota et al. 2011). The model involves three novel biological pathways: HD-Zip III genes, cytokinin network, and the Argonaute family members. During sepal development, IPT5 is activated by REV, while *IPT5* gene downregulation can be induced by the loss of AS1 gene function. REV protein, an HD-Zip Class III member, regulates sepal development through the following pathways: (1) REV protein activates AS1, which is mediated by IPT5, CK, and GTE6; (2) REV activates TAS3 siRNA, which is mediated by AGO7; (3) REV inhibits the activity of AGO1 through AGO10.

7.5.4 HD-Zip Class IV

HD-Zip IV genes are specifically expressed in epidermal cells of plants. They mainly regulate epidermal cell differentiation, root growth, anthocyanin accumulation, and trichome formation (Henriksson et al. 2005).

Maize OCL1 is an HD-Zip Class IV member, and its activity relies on the 85 amino acids at the N-terminus of the START domain. Transgenic OCL1-overexpressing maize plants showed various phenotypic defects accompanied by delayed flowering (Depege-Fargeix et al. 2011). Maize OCL4 is involved in anther cell differentiation. AMHDG1 in *Ammopiptanthus mongolica* is an HD-Zip IV protein. *Arabidopsis* plants with heterologous overexpression of AMHDG1 showed various phonotypes, including noncurling leaves and nondehiscent anthers. These plants displayed a downregulation of *AtNST1* and *AtNST2*, which are two positive regulators of anther dehiscence. These results suggest that AMHDG1 is a negative regulator of anther dehiscence (Wei et al. 2012).

OsROC5 and OsROC8 are HD-ZIP IV family transcription factors that form a transcriptional repressor complex with the transcriptional repressor cofactor TPL2, directly binding to the OsACL1 promoter, a positive regulator of vesicular cell development, and inhibiting its expression, thereby regulating leaf development (Fang et al. 2021). In *OsROC5* and *OsROC8* genes, the number and size increased in the absence of OsROC5 and *OsROC8* genes, the number and size of the leaves and the wild type in plant height, spike type, fertility, and grain size (Sun et al. 2020).

7.6 Prospects

As shown by the research on the structure and function of HD-Zip transcription factors using the model plants *Arabidopsis* and rice, despite their high phylogenetic similarity, the HD-Zip subfamily members have different spatiotemporal expression patterns. The members play different regulatory roles in response to the same environmental stress factors and function redundantly in the regulation of plant growth and development. Therefore, the biological function of an HD-Zip family member in a non-model species can be deduced from the alignment of sequences and conserved domains after extraction. This process may also lead to the discovery of species-specific genetic resources. Through integrating exogenous environmental signals and endogenous hormone signals and participating in signal cross-interactions to mediate downstream gene expression, HD-Zip transcription factors regulate plant growth and development and promote the adaptive growth of plants under adverse conditions. However, the regulatory network mechanism of HD-Zip remains to be further clarified, e.g., how HD-Zip transcription factors participate in signaling pathways under adverse environmental conditions and how HD-Zip transcription factors control the downstream genes. Further in-depth research on identification, alignment analysis, and function analysis of homologous genes in different species can deepen our understanding of HD-Zip transcription factors and help us reveal the molecular mechanisms underlying their regulatory effect on normal growth and development and on the adaptive growth of plants under adverse conditions. The development of bioinformatics and molecular biology will undoubtedly promote research on HD-Zip transcription factors.

References

Abe, M., Takahashi, T., and Komeda, Y. (2001). Identification of a cis-regulatory element for L1 layer-specific gene expression, which is targeted by an L1-specific homeodomain protein. *Plant J* 26: 487–494.

Agalou, A., Purwantomo, S., Overnäs, E. et al. (2008). A genome-wide survey of HD-Zip genes in rice and analysis of drought-responsive family members. *Plant Mol Biol* 66: 87–103.

Ahmad, S., Chen, Y., Shah, A.Z. et al. (2022). The Homeodomain-Leucine zipper genes family regulates the Jinggangmycin mediated immune response of Oryza sativa to Nilaparvata lugens, and Laodelphax striatellus. *Bioengineering* 9: 398.

An, P., Qin, R., Zhao, Q. et al. (2022). Genetic transformation of LoHDZ2 and analysis of its function to enhance stress resistance in Larix olgensis. *Sci Rep* 12: 12831.

Aoyama, T., Dong, C.H., Wu, Y. et al. (1995). Ectopic expression of the Arabidopsis transcriptional activator Athb-1 alters leaf cell fate in tobacco. *Plant Cell* 7: 1773–1785.

Ariel, F.D., Manavella, P.A., Dezar, C.A., and Chan, R.L. (2007). The true story of the HD-Zip family. *Trends Plant Sci* 12: 419–426.

Baima, S., Possenti, M., Matteucci, A. et al. (2001). The Arabidopsis ATHB-8 HD-zip protein acts as a differentiation-promoting transcription factor of the vascular meristems. *Plant Physiol* 126: 643–655.

Barrero, J.M., Millar, A.A., Griffiths, J. et al. (2010). Gene expression profiling identifies two regulatory genes controlling dormancy and ABA sensitivity in Arabidopsis seeds. *Plant J* 61: 611–622.

Bayarsaihan, D., Enkhmandakh, B., Makeyev, A. et al. (2003). Homez, a homeobox leucine zipper gene specific to the vertebrate lineage. *Proc Natl Acad Sci* 100: 10358–10363.

Brady, S.M., Sarkar, S.F., Bonetta, D., and McCourt, P. (2003). The ABSCISIC ACID INSENSITIVE 3 (ABI3) gene is modulated by farnesylation and is involved in auxin signaling and lateral root development in Arabidopsis. *Plant J* 34: 67–75.

Cabello, J.V., Arce, A.L., and Chan, R.L. (2012). The homologous HD-Zip I transcription factors HaHB1 and AtHB13 confer cold tolerance via the induction of pathogenesis-related and glucanase proteins. *Plant J* 69: 141–153.

Carabelli, M., Morelli, G., Whitelam, G., and Ruberti, I. (1996). Twilightzone and canopy shade induction of the Athb-2 homeobox gene in green plants. *Proc Natl Acad Sci USA* 93: 3530–3535.

Carlsbecker, A., Lee, J.Y., Roberts, C.J. et al. (2010). Cell signaling by microRNA165/6 directs gene dose-dependent root cell fate. *Nature* 465: 316–321.

Chan, R.L., Gago, G.M., Palena, C.M., and Gonzalez, D.H. (1998). Homeoboxes in plant development. *Biochim Biophys Acta* 1442: 1–19.

Ciarbelli, A., Ciolfi, A., Salvucci, S. et al. (2008). The Arabidopsis Homeodomain-leucine Zipper II gene family: diversity and redundancy. *Plant Mol Biol* 68: 465–478.

Deng, X., Phillips, J., Bräutigam, A. et al. (2006). A homeodomain leucine zipper gene from Craterostigma plantagineum regulates abscisic acid responsive gene expression and physiological responses. *Plant Mol Biol* 61: 469–489.

Depege-Fargeix, N., Javelle, M., Chambrier, P. et al. (2011). Functional characterization of the HD-ZIP IV transcription factor OCL1 from maize. *J Exp Bot* 62: 293–305.

Dezar, C.A., Gago, G.M., Gonzalez, D.H., and Chan, R.L. (2005). Hahb-4, a sunflower homeobox-leucine zipper gene, is a developmental regulator and confers drought tolerance to Arabidopsis thaliana plants. *Transgenic Res* 4: 429–440.

Dezar, C.A., Giacomelli, J.I., Manavella, P.A. et al. (2011). HAHB10, a sunflower HD-Zip II transcription factor, participates in the induction of flowering and in the control of phytohormone-mediated responses to biotic stress. *J Exp Bot* 62: 1061–1076.

Du, J., Miura, E., Robischon, M. et al. (2011). The Populus Class III HD ZIP transcription factor POPCORONA affects cell differentiation during secondary growth of woody stems. *PLoS One* 6: e17458.

Elhiti, M. and Stasolla, C. (2009). Structure and function of homodomain-leucine zipper (HD-Zip) proteins. *Plant Signal Behav* 4: 86–88.

Fang, J.J., Guo, T.T., Xie, Z.W. et al. (2021). The URL1-ROC5-TPL2 transcriptional repressor complex represses the ACL1 gene to modulate leaf rolling in rice. *Plant Physiol* 185: 1722–1744.

Hawker, N.P. and Bowman, J.L. (2004). Roles for Class III HD-Zip and KANADI genes in Arabidopsis root development. *Plant Physiol* 135: 2261–2270.

Henrik, J., Yan, W., Johannes, H., and Peter, E. (2003). The Arabidopsis thaliana homeobox gene ATHB5 is a potential regulator of abscisic acid responsiveness in developing seedlings. *Plant Mol Biol* 51: 719–729.

Henriksson, E., Olsson, A.S., Johannesson, H. et al. (2005). Homeodomain leucine zipper class I genes in Arabidopsis. Expression patterns and phylogenetic relationships. *Plant Physiol* 139: 509–518.

Himmelbach, A., Hoffmann, T., Leube, M. et al. (2002). Homeodomain protein ATHB6 is a target of the protein phosphatase ABI1 and regulates hormone responses in Arabidopsis. *EMBO J* 21: 3029–3038.

Jiao, P., Jiang, Z., Wei, X. et al. (2022). Overexpression of the homeobox-leucine zipper protein ATHB-6 improves the drought tolerance of maize (Zea mays L.). *Plant Science* 316: 111159.

Johannesson, H., Wang, Y.A., and Engström, P. (2001). DNA-binding and dimerization preferences of Arabidopsis homeodomain-leucine zipper transcription factors in vitro. *Plant Mol Biol* 45: 63–73.

Kang, J. and Dengler, N. (2002). Cell cycling frequency and expression of the homeobox gene ATHB-8 during leaf vein development in Arabidopsis. *Planta* 216: 212–219.

La Rota, C., Chopard, J., Das, P. et al. (2011). A data-driven integrative model of sepal primordium polarity in Arabidopsis. *Plant Cell* 23: 4318–4333.

Lechner, E., Leonhardt, N., Eisler, H. et al. (2011). MATH/BTB CRL3 receptors target the homeodomain-leucine zipper ATHB6 to modulate abscisic acid signaling. *Dev Cell* 21: 1116–1128.

Lee, Y.H., Oh, H.S., Cheon, C.I. et al. (2001). Structure and expression of the Arabidopsis thaliana homeobox gene Athb-12. *Biochem Biophys Res Commun* 284: 133–141.

Li, X., Hou, Y., Zhang, F. et al. (2022). Identification and characterization of stress responsive homeodomain leucine zipper transcription factors in Medicago truncatula. *Mol Biol Rep* 49: 3569–3581.

Lin, Z., Hong, Y., Yin, M. et al. (2008). A tomato HD-Zip homeobox protein, LeHB-1, plays an important role in floral organogenesis and ripening. *Plant J* 55: 301–310.

Liu, X., Cheng, L., Li, R. et al. (2022). The HD-Zip transcription factor SlHB15A regulates abscission by modulating jasmonoyl-isoleucine biosynthesis. *Plant Physiology* 189: 2396–2412.

Manavella, P.A., Arce, A.L., Dezar, C.A. et al. (2006). Cross-talk between ethylene and drought signalling pathways is mediated by the sunflower Hahb-4 transcription factor. *Plant J* 48: 125–137.

Manavella, P.A., Dezar, C.A., Bonaventure, G. et al. (2008). HAHB4, a sunflower HD-Zip protein, integrates signals from the jasmonic acid and ethylene pathways during wounding and biotic stress responses. *Plant J* 56: 376–388.

McConnell, J.R. and Barton, M.K. (1998). Leaf polarity and meristem formation in Arabidopsis. *Development* 125: 2935–2942.

McConnell, J.R., Emery, J., Eshed, Y. et al. (2001). Role of PHABULOSA and PHAVOLUTA in determining radial patterning in shoots. *Nature* 411: 709–713.

McIntosh, B.E., Hogenesch, J.B., and Bradfield, C.A. (2010). Mammalian Per-Arnt-Sim proteins in environmental adaptation. *Annu Rev Physiol* 72: 625–645.

Möglich, A., Ayers, R.A., and Moffat, K. (2009). Structure and signaling mechanism of Per-ARNT-Sim domains. *Structure* 17: 1282–1294.

Mukherjee, K. and Bürglin, T.R. (2006). MEKHLA, a novel domain with similarity to PAS domains, is fused to plant homeodomain-leucine zipper III proteins. *Plant Physiol* 140: 1142–1150.

Nakamura, M., Katsumata, H., Ab, M. et al. (2006). Characterization of the class IV homeodomain-Leucine Zipper gene family in Arabidopsis. *Plant Physiol* 141: 1363–1375.

Ochando, I., Jover-Gil, S., Ripoll, J.J. et al. (2006). Mutations in the microRNA complementarity site of the INCURVATA4 gene perturb meristem function and adaxialize lateral organs in Arabidopsis. *Plant Physiol* 141: 607–619.

Ohashi-Ito, K. and Fukuda, H. (2003). HD-Zip III homeobox genes that include a novel member, ZeHB-13 (Zinnia)/ATHB-15 (Arabidopsis), are involved in procambium and xylem cell differentiation. *Plant Cell Physiol* 44: 1350–1358.

Otsuga, D., DeGuzman, B., Prigge, M.J. et al. (2001). REVOLUTA regulates meristem initiation at lateral positions. *Plant J* 25: 223–236.

Palena, C.M., Gonzalez, D.H., and Chan, R.L. (1999). A monomer-dimer equilibrium modulates the interaction of the sunflower homeodomain leucine-zipper protein Hahb-4 with DNA. *Biochem J* 341: 81–87.

Palena, C.M., Tron, A.E., Bertoncini, C.W. et al. (2001). Positively charged residues at the N-terminal arm of the homeodomain are required for efficient DNA binding by homeodomain-leucine zipper proteins. *J Mol Biol* 308: 39–47.

Peterson, K.M., Shyu, C., Burr, C.A. et al. (2013). Arabidopsis homeodomain-leucine zipper IV proteins promote stomatal development and ectopically induce stomata beyond the epidermis. *Development* 140: 1924–1935.

Prigge, M.J., Otsuga, D., Alonso, J.M. et al. (2005). Class III homeodomain-leucine zipper gene family members have overlapping, antagonistic, and distinct roles in Arabidopsis development. *Plant Cell* 17: 61–76.

Qiu, X., Wang, G., Abou-Elwafa, S.F. et al. (2022). Genome-wide identification of HD-ZIP transcription factors in maize and their regulatory roles in promoting drought tolerance. *Physiol Mol Biol Plants* 28: 425–437.

Ré, D.A., Dezar, C.A., Chan, R.L. et al. (2011). Nicotiana attenuata NaHD20 plays a role in leaf ABA accumulation during water stress, benzylacetone emission from flowers, and the timing of bolting and flower transitions. *J Exp Bot* 62: 155–166.

Robischon, M., Du, J., Miura, E., and Groover, A. (2011). The Populus class III HD ZIP, popREVOLUTA, influences cambium initiation and patterning of woody stems. *Plant Physiol* 155: 1214–1225.

Rueda, E.C., Dezar, C.A., Gonzalez, D.H., and Chan, R.L. (2005). Hahb-10, a sunflower homeobox-leucine zipper gene, is regulated by light quality and quantity, and promotes early flowering when expressed in Arabidopsis. *Plant Cell Physiol* 46: 1954–1963.

Shaikhali, J., Norén, L., de Dios Barajas-López, J. et al. (2012). Redox-mediated mechanisms regulate DNA binding activity of the G-group of basic region leucine zipper (bZIP) transcription factors in Arabidopsis. *J Biol Chem* 287: 27510–27525.

Söderman, E., Hjellstrom, M., Fahleson, J., and Engstrom, P. (1999). The HD-Zip gene ATHB6 in Arabidopsis is expressed in developing leaves, roots and carpels and up-regulated by water deficit conditions. *Plant Mol Biol* 40: 1073–1083.

Söderman, E., Mattsson, J., and Engström, P. (1996). The Arabidopsis homeobox gene ATHB-7 is induced by water deficit and by abscisic acid. *Plant J* 10: 375–381.

Son, O., Hur, Y.S., Kim, Y.K. et al. (2010). ATHB12, an ABA-inducible homeodomain-leucine zipper (HD-Zip) protein of Arabidopsis, negatively regulates the growth of the inflorescence stem by decreasing the expression of a gibberellin 20-oxidase gene. *Plant Cell Physiol* 51: 1537–1547.

Sorin, C., Salla-Martret, M., Bou-Torrent, J. et al. (2009). ATHB4, a regulator of shade avoidance, modulates hormone response in Arabidopsis seedlings. *Plant J* 59: 266–277.

Stamm, P. and Kumar, P.P. (2010). The phytohormone signal network regulating elongation growth during shade avoidance. *J Exp Bot* 61: 2889–2903.

Steindler, C., Carabelli, M., Borello, U. et al. (1997). Phytochrome A, phytochrome B and other phytochrome(s) regulate ATHB-2 gene expression in etiolated and green Arabidopsis plants. *Plant Cell Environ* 20: 759–763.

Sun, J., Cui, X.A., Teng, S.Z. et al. (2020). HD-ZIP IV gene Roc8 regulates the size of bulliform cells and lignin content in rice. *Plant Biotechnol J* 18 (12): 2559–2572.

Talbert, P.B., Adler, H.T., Parks, D.W., and Comai, L. (1995). The REVOLUTA gene is necessary for apical meristem development and for limiting cell divisions in the leaves and stems of Arabidopsis thaliana. *Development* 121: 2723–2735.

Tron, A.E., Bertoncini, C.W., Chan, R.L., and Gonzalez, D.H. (2002). Redox regulation of plant homeodomain transcription factors. *J Biol Chem* 277: 34800–34807.

Tron, A.E., Bertoncini, C.W., Palena, C.M. et al. (2001). Combinatorial interactions of two amino acids with a single base pair define target site specificity in plant dimeric homeodomain proteins. *Nucleic Acids Res* 29: 4866–4872.

Tron, A.E., Comelli, R.N., and Gonzalez, D.H. (2005). Structure of homeodomain-leucine zipper/DNA complexes studied using hydroxyl radical cleavage of DNA and methylation interference. *Biochemistry* 44: 16796–16803.

Valdés, A.E., Övernäs, E., Johansson, H. et al. (2012). The homeodomain-leucine zipper (HD-Zip) class I transcription factors ATHB7 and ATHB12 modulate abscisic acid signalling by regulating protein phosphatase 2C and abscisic acid receptor gene activities. *Plant Mol Biol* 80: 405–418.

Wang, K., Xu, L., Wang, Y. et al. (2022b). Genome-wide characterization of homeodomain-leucine zipper genes reveals RsHDZ17 enhances the heat tolerance in radish (Raphanus sativus L.). *Physiol Plant* 174: e13789.

Wang, Y., Henriksson, E., Söderman, E. et al. (2003). The Arabidopsis homeobox gene, ATHB16, regulates leaf development and the sensitivity to photoperiod in Arabidopsis. *Dev Biol* 264: 228–239.

Wei, Q., Kuai, B.K., Hu, P., and Ding, Y.L. (2012). Ectopic-overexpression of an HD-Zip IV transcription factor from Ammopiptanthus mongolicus (Leguminosae) promoted upward leaf curvature and non-dehiscent anthers in Arabidopsis thaliana. *Plant Cell Tiss Org Cult* 110: 299–306.

Williams, L., Grigg, S.P., Xie, M.T. et al. (2005). Regulation of Arabidopsis shoot apical meristem and lateral organ formation by microRNA miR166g and its AtHD-ZIP target genes. *Development* 132: 3657–3668.

Yang, Q., Xiang, W., Li, Z. et al. (2022). Genome-wide characterization and expression analysis of HD-ZIP gene family in dendrobium officinale. *Front Genet* 13: 797014.

Yang, Y.F., Al-Baidhani, H.H.J., Harris, J. et al. (2020). DREB/CBF expression in wheat and barley using the stress-inducible promoters of HD-Zip I genes: impact on plant development, stress tolerance and yield. *Plant Biotechnol J* 18 (3): 829–844.

Zhang, T., You, J., Zhang, Y. et al. (2021) LF1 regulates the lateral organs polarity development in rice. New Phytol 231: 1265–1277.

Zhong, R.Q. and Ye, Z.H. (1999). IFL1, a gene regulating interfascicular fiber differentiation in Arabidopsis, encodes a homeodomain-leucine zipper protein. *Plant Cell* 11: 2139–2152.

Zhong, R.Q. and Ye, Z.H. (2004). Amphivasal vascular bundle 1, a gain-of-function mutation of the IFL1/REV gene, is associated with alterations in the polarity of leaves, stems and carpels. *Plant Cell Physiol* 45: 369–385.

Zhu, Y., Peng, S., Zhao, L. et al. (2022). Genome-wide identification and characterization of the HD-Zip gene family and expression analysis in response to stress in Rehmannia glutinosa Libosch. *Plant Signal. Behav.* 17: 2096787.

8

SBP-box Transcription Factors and Flesh Fruit Development and Ripening

8.1 Research Progress on Plant SBP Family Transcription Factors

Research on transcription factors has been conducted in many organisms, and important discoveries have been made. For example, transcription factors in plants are involved in various biological processes, including responses to environmental stresses, regulation of plant defense against pathogenic microorganisms, plant growth regulation, formation of some special organs (such as flowers), differentiation, and metabolism. A large number of transcription factor families, including *MADS*, *AP2/ERF*, *EIN3/EILS*, and *SBP*, have been identified in plants. Among them, the SBP-box gene family is believed to be specific to plants. In a study in which two founding members, *SBP1* and *SBP2*, were isolated from snapdragon (*Antirrhinum majus*), the researchers found that early floral development could be controlled by regulation of the MADS-box gene *SQUAMOSA* (Klein et al. 1996). "*SQUAMOSA* promoter-binding protein" (SBP) was so named because this protein was shown to recognize and specifically bind to the promoter of *SQUA* gene in *Antirrhinum majus*. Later, SBP-like (*SPL*) genes were identified in various plants, including green algae (*Chlamydomonas reinhardtii*), moss, silver birch (*Betula platyphylla*), tomato (*Solanum lycopersicum*), rice (*Oryza sativa*), and maize (*Zea mays*) (Moreno et al. 1997; Lannenpaa et al. 2004). Since their discovery, SBP family transcription factors have gradually become a research hotspot in plant molecular biology. As an increasing number of *SBP* genes have been identified in plants and transcriptome sequencing technologies have been improved in recent years, SBP-box genes have been identified and analyzed in many species by using bioinformatic methods. These genes have various important biological functions, including sporogenesis, branch development, floral development, leaf development, and responses to environmental stresses (Arazi et al. 2005; Manning et al. 2006).

Transcriptional Regulation of Flesh Fruit Development and Ripening, First Edition. Yunbo Luo.

8.1.1 Origin and Development of the SBP-box Gene Family

SBP was initially identified in *Antirrhinum majus*. In a study investigating genes involved in the regulation of plant floral organ formation, Huijser et al. found that nucleoproteins have several binding sites to the promoter of the *Antirrhinum majus* meristem identity gene *SQUA* in an electrophoretic mobility shift assay (EMSA). Subsequently, two related genes were detected by screening the inflorescence cDNA expression library of *Antirrhinum majus*. The amino acid sequences encoded by the two genes contained a highly conserved DNA-binding domain consisting of approximately 80 amino acid residues and were produced before expression of the *SQUA* gene. This finding indicates that the nucleoproteins identified above may specifically bind to the promoter of the meristem identity gene and activate gene transcription in *Antirrhinum majus*, suggesting that they are new transcription factors regulating early floral development (Klein and Huijser 1996).

8.1.2 Structure of SBP Family Transcription Factors

All SBP family transcription factors contain a conserved DNA-binding domain (SBP-DBD) consisting of 76 amino acid residues. The binding domain of SBP-DBD has three antiparallel β-sheets at the N-terminus and one or two short helices at the C-terminus. The backbone pairwise root mean-square deviation (RMSD) value of the distance between the secondary structure elements packed in a similar manner among SBPs is 1.86 A° (Kazuhiko et al. 2004). This finding gives rise to speculation that SBP-DBD contains two subdomains that are packed tightly to each other through interactions of hydrophobic residues and that each subdomain contains a binding pocket fixed by the bond required for binding with zinc ions. Because these hydrophobic residues are highly conserved among domains, packing of the two subdomains is likely to be necessary to maintain the structure, allowing SBP-DBD to behave essentially as a single domain.

The size of *Arabidopsis thaliana* (At) SPL varies significantly, ranging from 131 amino acids (AtSPL3) to 927 amino acids (AtSPL12) (Cardon et al. 1999). In 2004, nuclear magnetic resonance (NMR) technology was used to analyze the three-dimensional solution conformation of two AtSBP domains (SPL4 and SPL7) and revealed that they both contain two zinc finger structures necessary for their functions and a nuclear localization signal (NLS). The *SPL* gene consists of a large number of exons, but only the first and second exons contribute to SBP encoding in plants (Guo, Song et al. 2008). The C-terminus of SBP contains a highly conserved serine (Ser) that might be located in the NLS region, which may be involved in protein phosphorylation. The zinc finger structure is composed of eight *Cys/His* residues, and every four amino acid residues bind one zinc ion. The C-terminus is *Cys-Cys-Cys-His* or *Cys-Cys-Cys-Cys*, and the N-terminus is *Cys-Cys-His-Cys*. This zinc finger structure of the *SBP* gene family is found only in plants, which is different from the common interleaved pattern where the 1st, 2nd, 5th, and 6th residues bind to one zinc ion and the 3rd, 4th, 7th, and 8th residues bind to another zinc ion. Although the researchers confirmed the presence of zinc fingers in SBPs, they did not address whether both zinc fingers contribute to DNA binding and whether the zinc finger structure affected the binding of a transcription factor to DNA (Kazuhiko Yamasaki et al. 2004). Later, an ethylene diamine tetraacetic acid (EDTA) chelation experiment targeting the SPL1-SBP region showed that the DNA-binding efficiency of an SBP remained unchanged in the absence of EDTA but was abolished by the presence of EDTA. The loss of the DNA-binding ability could be rescued by adding Zn^{2+} but not Mg^{2+} after urea renaturation. This finding demonstrates that Zn^{2+} plays an important role in the binding of SBP to DNA (Birkenbihl et al. 2005).

In the initial study that led to the discovery of SBP transcription factors, the researchers speculated the existence of the zinc finger structure in SBP but did not confirm the effect of the zinc finger structure on the binding efficiency of SPL12 to the *SQUA* sequence in EDTA titration tests. Later, the researchers found that the SBP region of SPL12 could maintain its structure even when it could only bind one Zn^{2+} due to a loss of a cysteine (Cys) at the C-terminus of the zinc finger structure. This discovery led to speculation that Zn^{2+} at the N-terminus is indispensable for the structure of SBP and that the second Zn^{2+} mainly plays the role of guiding the C-terminal loop of SBP to position properly into the major groove of the DNA for DNA binding (Kazuhiko et al. 2006). Therefore, the two zinc fingers in SBP-DBD have different functions, and both contribute to the function of transcription factors.

From the structure of the SBP family members found thus far in various plants, it is clear that they all bind to DNA in a highly conservative manner. It is generally believed that the promoter sequence of the gene encoding SBP-DBD is *TNCGTACAA*, where N represents any base (Cardon et al. 1997). The region of the protein surface responsible for DNA binding is charged positively, and the DNA is negatively charged. This region in the SBP-DBD has a continuous area of the surface that is positively charged and includes all the conserved basic residues, and therefore, it is highly likely to be involved in DNA binding. Helices are found in both the N-terminal and C-terminal regions of SPL-DBD. At the N-terminus, the Arg/Lys sidechains protruding from the surface near the N-terminal α-helices and most of the DBDs can recognize DNA bases through α-helices. Histidine (His) and asparagine (Asp) are among the typical base-recognition amino acid residues, which suggests the possibility that at least a part of the C-terminal region forms a helix in the complex with DNA and the possibility that some residues in this region, together with the N-terminal helix, recognize DNA bases to achieve sequence specificity. The positively charged region extending from the N-terminus to the C-terminus of SPL-DBD has a saddle-shaped surface, forming a concave area parallel to the major groove of the double-stranded DNA. Meanwhile, the surface of the major groove of the DNA is convex along the groove extending helically, forming convex areas perpendicular to the concave area of SPL-DBD. Hence, the surface of the positively charged region of the SPL4-DBD is sufficiently complementary to that of the major groove of the DNA. Furthermore, the flexibility in the C-terminal region enables elongation of the contact area, allowing a close fit with the DNA. The contact area of SPL4-DBD covers a region of approximately 96 bp, including the known recognition sequence, TNCGTACAA. In addition, the model proposed by Yamasaki et al. suggested that the N-terminal helix and a part of the C-terminal region that can recognize sidechains of several residues in the major groove are positioned inside the major groove to contact bases in the model. However, further fitting of the amino acid sidechains and bases is required to achieve the specificity to the target sequence. Moreover, there is an NLS at the C-terminus of SBP, which partially overlaps with the zinc finger structure of the SBP-DBD. Through transient expression of SPL3-green fluorescent protein (GFP) fusion constructs and SPL3 immunolocalization, Birkenbihl et al. found that the SBP domain protein LIGULELESS (LG1) could still be imported into the nucleus when the SBP-domain was removed. The amino acid exchange experiment confirmed that the function of the SBP-domain was regulated by posttranslational modification, which affected its ability to bind DNA and enter the nucleus. The researchers also found that the palindromic GTAC core motif of the DNA is essential for efficient DNA binding with SBP transcription factors. This finding confirmed the biological function of NLS. Protein sequence analysis revealed that AtSBP with a short amino acid sequence generally had only one NLS while those with more than 1000 amino acid residues might contain some other NLSs. NLSs could guide the nuclear importation of SBP to regulate the transcriptional expression of its downstream genes. This conserved structure of the *SBP* gene family is present in different plant species (Birkenbihl et al. 2005).

In summary, the *SBP* gene family members contain a highly conserved sequence consisting of 76 amino acid residues. They all have two zinc fingers and one NLS. The zinc-coordinating pattern of the zinc finger structure, which is plant-specific, is not the common interleaved pattern. The N-terminal Zn^{2+} and NLS are essential for transcription factors to bind DNA.

8.1.3 Evolutionary Analysis of the *SBP* Gene Family

SBP transcription factors have become a research hotspot since they were discovered. With the research advancement in molecular biology, an increasing number of plant *SBP* genes have been discovered, and a large amount of research has been directed toward investigating the evolutionary pattern of the *SBP* family in plants. As a result, phylogenetic trees based on the conserved sequences in the SBP region have been established for some plant species. By identifying 120 SBP-box genes in algae, bryophytes, monocotyledons, dicotyledons, and other plants, researchers have constructed a phylogenetic tree based on the maximum likelihood method and proposed an evolutionary model of the SBP-box gene family.

SBP-box genes emerged prior to the divergence of the common ancestor shared by the green alga *C. reinhardtii* and terrestrial plants. As shown by structural and phylogenic analysis of exons and introns in SBP-box genes of terrestrial plants, divergence of SBP-box genes in terrestrial plants occurred after the divergence of the most recent common ancestor shared by green algae and terrestrial plants. Therefore, the SBP-box genes of green algae can be classified into a single clade (i.e., the CR group). With the evolution of terrestrial plants, SBP-box genes underwent some changes, and as a result, SBP-box genes are divided into group I and group II according to their structure. Researchers have speculated that the SBP ancestor gene of terrestrial plants might have a complex gene structure with many exons and two zinc fingers, C3H and C2HC, and the ancestor gene then diverged into groups I and II. Among them, the group I ancestor gene has evolved into the group I genes by altering the first zinc finger from C3H to C4. The group II ancestor gene might have duplicated into three copies, which have evolved into the subgroups IIa, IIb, and IIc-Ig. Group I includes six SBP-box genes in five plant species, including two different genes in poplar and one gene identified from each of four other plant species. This grouping is different from the constitution of other groups, in which the SBP-box gene members from the same plant species are mostly clustered into a single group. Subgroups IIa and IIb contain only the SBP-box genes from terrestrial plants. Nearly half of moss and lycophyte SBP-box genes belong to subgroup IIb, which also includes one *Arabidopsis* and three rice genes. The SBP-box genes in subgroup IIc are all from vascular plants, and those of subgroup IId are from seed plants. All three genes in group IIg are from moss. In addition, the members of subgroups IIc-IIg all have a conserved miRNA target site in their exons or 3′-UTR, suggesting that the miRNA target site emerged prior to the divergence of the lineages. The evolutionary model described above clearly demonstrates the evolutionary process of the SBP-Box gene family; that is, all *SBP* genes in green algae are grouped into one group (the CR cluster), and those from other terrestrial plants are grouped into two different groups. This arrangement indicates that the presence of *SBP* genes might have been earlier than the divergence between green algae and terrestrial plants. *SBP* genes may have originated from a common ancestor of green plants, and then duplication and divergence in each lineage including exon–intron loss processes have led to the gradual changes in the genes (Guo, Zhu et al. 2008).

8.1.4 Regulation of the SBP-box Gene Family

MicroRNAs (miRNAs) are a kind of endogenous noncoding RNA with regulatory functions (Jeena et al. 2022). MiRNAs are found in eukaryotes and have a size of approximately 20–25 nucleotides (Samad et al. 2016). MiRNAs are important transcriptional regulators that function mainly through degrading mRNA of target genes and inhibiting their transcriptional translation (Bartel 2004). Current research on the expression regulation of the SBP-Box gene family mainly focuses on miRNA regulation. Mature miRNA is generated through nuclease cleavage of primary miRNA (pri-miRNA), which incorporates into the RNA-induced silencing complex (RISC). By recognizing target mRNA through base-pairing, the miRNA guides RISC to degrade the target mRNA or impede its translation depending on the degree of complementarity. Recent studies have shown that miRNA is involved in a variety of regulatory pathways, including development, organogenesis, hematopoiesis, viral defense, lipid metabolism, cell proliferation, and apoptosis (Wang et al. 2016).

MiR156 is a highly conserved miRNA gene family that exists in higher plants such as moss, lycophytes, ferns, monocotyledons, and dicotyledons. Previous studies have confirmed that miR156/157 target *SBP* genes. At present, 11 of the 17 SBP transcription factors found in *A. thaliana* contain the regulatory sites targeted by miR156/157 (Rhoades et al. 2002). Schwab et al. reported that miR156 regulated *SPL3* expression at the stage of vegetative development. In *A. thaliana*, the *SPL* gene targeted by miR156 has dual effects on plastochron length and organ size. MiR156 quantitatively regulates *SPL* expression in leaf primordia and inhibits the formation of new leaves at the shoot apical meristem. The effects of miR156-targeted SPL genes on organ size and changes in plastochron length are correlated with CYP78A5 (Wang et al. 2008). In *A. thaliana*, miR172 signaling participates in the signal pathway of miR156 regulating *SPL3*, *SPL4*, and *SPL5*. *SBP3* in moss also contains a regulatory sequence targeted by miR156. The 3′UTR end of *SPL3*, *SPL4*, and *SPL5* contains a regulatory sequence complimentary to miR156, and therefore, miR156 is involved in the transition from the juvenile to adult phase by regulating the expression of *SPL3*. Hence, reducing the amount of miR156 can enhance *SPL3* expression and promote early flowering of plants (Gandikota et al. 2007b). In addition, the 3′UTR end of *SPL3* contains the regulatory sequence complementary to miR156 and miR157, named the miRNA responsive element (MRE). In transgenic plants constitutively overexpressing an *SPL3* transgene, the altered MRE sequence did not affect the level of *SPL3* transcripts, while the plants without the altered MRE sequence did not have a detectable expression level of SPL3 protein, suggesting that miR156 inhibits the expression of *SPL3* posttranscriptionally (Gandikota et al. 2007b). Eleven *OsSPL* genes have been predicted to be target genes of *OsmiR156* in the *Oryza sativa* genome (Din et al. 2022). Similar to *A. thaliana*, most MREs are located within the coding region, with the exceptions of *OsSPL4* and *OsSPL13*, which contain target sites in the 3′UTR (Xie et al. 2006). Sequence alignment of 19 *OsSPL* genes and 12 *OsmiR156* precursors revealed that 11 *OsSPL* genes have regulatory sites complimentary to miR156; in other words, the 11 *OsSPL* genes are targets of *OsmiR156*. *OsmiR156* genes are highly expressed in young shoots and leaves of rice, resulting in severe dwarfism, strongly reduced panicle size, and delayed flowering. These phenotypes suggest that *OsmiR156* and its targeted *OsSPL* genes are involved in various developmental processes of rice, especially floral development. Ten of the 15 SBP family transcription factors identified in the tomato genome have miR156/157-binding elements (Lee et al. 2007). Analysis of data from the miRNA database and the tomato genome database identified eight miR156 and four miR157 genes. Tissue-specific expression analysis revealed that miR156 and miR157 are mainly expressed in seedlings, leaves, and anthers of young flowers, demonstrating an expression profile complementary to that of their targeted *SBP* gene (Jiao et al. 2010). MiR156 regulates *SPL9* expression primarily in a quantitative manner rather than a spatial manner. A large number of studies have shown that most members of the *SBP-box* gene family contain miRNA regulatory sequences with varying regulatory patterns; however, the underlying regulatory mechanism needs further in-depth study.

8.1.5 Biological Functions *of the SBP-box Gene* Family

8.1.5.1 Role of SBP Transcription Factors in Environmental Signal Response

Since their discovery, SBP transcription factors have gradually become a research hotspot in plant molecular biology, in particular in *A. thaliana*. Sixteen *AtSBP* members have been identified (Cardon et al. 1999). *AtSPL4* and *AtSPL7* genes have the NMR solution structure of DBDs. *AtSPL7* is involved in the regulation of copper ion balance and is a homolog of copper response regulator 1 (*CRR1*) in *C. reinhardtii* (Kazuhiko Yamasaki et al. 2004). *AtSPL7* could directly bind to the *GTAC* sequence of the miR398 promoter in vitro. In addition, *AtSPL7* was demonstrated to be crucial for the expression of miR397, miR408, and miR857, which are small molecules regulating copper homeostasis, and *AtSPL7* also acts as a chaperone in encoding copper transporters. The *AtSPL7* mutants showed a phenotype of impaired growth under low-copper conditions, suggesting that *AtSPL7* plays a very important role in the response to copper deficiency in *A. thaliana* (Yamasaki et al. 2009). In *A. thaliana*, the SBP-box gene can respond to various biological and abiotic stresses by interacting with genes involved in defense response pathways, and it also plays an important role in hormone-mediated signal transduction (Wang et al. 2009b). Fumonisin B1 is a fungal toxin that disrupts sphingolipid metabolic pathways by inhibiting ceramide synthase activity and thereby induces programmed cell death in plants. Existing studies have shown that the *AtSPL14* gene is involved in programmed cell death and plays a pivotal role in the sensitivity to Fumonisin B1 (Miura K and suoka M 2010). A similar function was also detected in the *SBP* gene *CRR1* in *C. reinhardtii* (Kropat et al. 2005). In *AtJMT*-overexpressing *Arabidopsis* plants, *AtSPL2* responds to the methyl jasmonate-mediated disease resistance pathway (Jung and Park 2007).

8.1.5.2 Roles of SBP Transcription Factors in Flower Formation and Development

In *Arabidopsis*, there are at least four floral inductive pathways, i.e., the autonomous pathway, gibberellin pathway, photoperiod pathway, and vernalization pathway. In apical meristems, these four pathways act in both an independent and overlapping manner and involve floral pathway integrators such as *SUPPRESSOR OF OVEREXPRESSION OF CON-STANS 1* (*SOC1*) and *FLOWERING LOCUS T* (*FT*) (Spanudakis and Jackson 2014). These integrators initiate flowering by activating the floral meristem identity genes *FRUITFULL* (*FUL*), *APETALA1* (*AP1*), and *LEAFY* (*LFY*) in *A. thaliana* (Shao et al. 1999).

For many genes related to floral development, the orthologs between *Arabidopsis* and other species have similar functions. Nevertheless, little is known about the function of upstream regulators of these flowering genes in species other than *A. thaliana*. In *Antirrhinum majus*, two *SBP* homologs, *SBP1* and *SBP2*, directly regulate the orthologous gene of *SQUAMOSA*, suggesting that these two genes play a key role before flower formation (Klein and Huijser 1996). *SBP1* and *SBP2* were the first identified *SBP* genes that can specifically bind to the floral development-related promoter *SQUAMOSA*. Later, these genes were also identified in other species. *SBP1* and *SBP2* genes are expressed earlier than *SQUAMOSA* genes during the vegetative growth stage, and the expression levels of these two genes begin to increase during inflorescence formation and development (Klein and Huijser 1996). Unlike the *SQUA* gene, the expression levels of *SBP1* and *SBP2* are extremely high in petals but very low in sepals, as shown by Northern blot analysis (Klein and Huijser 1996). *SBP1* and *SBP2* are expressed in both inflorescences and floral meristems, but the expression of the SQ*UA* gene was inhibited in floral meristem. Based on their spatiotemporal expression patterns and similar binding patterns, it is speculated that the *SBP1* and *SBP2* genes may be functionally redundant in the late stage of floral development, and the functional loss of these two genes may lead to delayed flowering (Klein and Huijser 1996). To further investigate the function of these two genes, nine *pTRV2-SBP1* strains were screened using reverse genetics through virus-mediated gene silencing. Among them, four did not flower in the incubator for six months, two died when they had respectively 23 and 35 leaves before flowering, and all the others showed significant flowering delay. All strains had more leaves than the control plants before flowering. The results suggest that *Am SBP1* plays an important role in the transition to flowering. *AtSPL3* contains a transcriptional activator domain. Constitutive overexpression of *AtSPL3* resulted in an early flowering phenotype. Using the constructed cDNA library, 12 members of *AtSBP* gene family were cloned. Further, Northern blot analysis showed that three genes, *AtSPL1*, *AtSPL2*, and *AtSPL12*, were constitutively expressed, and the other nine genes were specifically expressed (Cardon et al. 1999).

According to the length of genes, as well as exon–intron structural and sequence similarity, 16 *Arabidopsis SBP* genes are divided into various subfamilies. The three genes with short sequences, *SPL3*, *SPL4*, and *SPL5*, are in the same group. They are expressed in floral organ-related tissues, and their expression is repressed by miR156/157 (Wang et al. 2009a). In general, *AtSPL3*, *AtSPL4*, and *AtSPL5* genes have the highest sequence similarity to *SBP1* and *SBP2* in *Antirrhinum majus*. Functional analysis has shown that these three genes promote early flowering of *A. thaliana* through photoperiodic and autonomous pathways. In the photoperiodic pathway, early flowering is mainly regulated by miR156 through repressing the *SPL* gene (Cardon G et al. 1997). *AtSPL3* and its homologs *AtSPL4* and *AtSPL5* are extremely highly expressed in cells at the floral primordium. Overexpression of *AtSPL3/AtSPL4/AtSPL5* could cause early flowering under long-day conditions, while overexpression of their repressive regulator miR156 led to the delay of flowering in *A. thaliana* (Cardon et al. 1997). In recent years, studies have found that *SPL3* expression can be induced by long-day conditions in wild *A. thaliana*, and the expression increase begins at the preflowering development stage. *AtSPL3* can directly bind the promoter regions of *AP1*, *FUL*, and *LFY*, and together with *AtSPL9*, it affects floral organ development in *A. thaliana* (Yamasaki et al. 2009). Leaf *SPL3* can promote flowering through indirectly interacting with the *FT* gene. In contrast, the expressed *SPL3*, *SPL4*, and *SPL5* in apical meristem can directly interact with *FUL* in inflorescence meristem, along with the floral meristem identity genes *AP1* and *LFY* (Wang et al. 2009a). In a study searching for the insertion sequence of the *SBP* gene in the transposon mutants of *A. thaliana*, three *AtSPL8* genes that are alleles of the transposon insertions were obtained and successfully characterized. *AtSPL8* knockout and mutation can affect micro- and megasporogenesis, formation of sepal epidermal hair, and extension of stamen filaments, suggesting that the function of *AtSPL8* is most similar to that of *LG1*. *AtSPL8*-deficient mutants showed a decline in reproductive capability due to abnormal anther cell differentiation, indicating that this gene is required for proper micro- and megasporogenesis. Research on *AtSPL8* gain- and loss-of-function mutants has shown that this gene is involved in gibberellin (GA)-mediated development; it promotes GA expression in flowers and acts as a negative regulator in seedlings (Zhang et al. 2007). All three independent transposon *SPL8* mutants successfully isolated from the transposon mutant populations had less pollen, smaller anthers, shorter pods, and reduced fertility. Histological analysis indicated that the reduced fertility was mainly associated with abnormal development of microsporangia. Knockout of *AtSPL8* also affected megasporogenesis, trichome formation of sepals, and extension of stamen filaments. Overexpression of *AtSPL8* in *AtSPL8*-deficient mutants could not restore the fertility of the mutants. Overexpression of *AtSPL8* in wild *A. thaliana* reduced its seed germination rate and fertility, which was demonstrated to be caused by anther indehiscence. These results suggest that *AtSPL8* plays an important role in the gibberellin-mediated anther developmental process (Zhang et al. 2007). In addition, the three *Arabidopsis SPL8/SPL9/SPL15*, *SPL8/SPL2/SPL9*, and *SPL8/SPL2/SPL15* mutant plants showed a lower fertility than that of single *SPL8* mutant plants. The *Arabidopsis SPL8* mutant plants became completely sterile after overexpression of the *SPL8* repressor miR156. All of the results above suggest that *SPL8* plays a key role in maintaining plant fertility in *Arabidopsis* (UnteUS et al. 2003).

In addition, SBP-DPD was detected in the promoter of *BpMADS5* (a MADS-box gene highly homologous to *AtFRUITFULL*) in silver birch (*B. platyphylla*), leading to the discovery and successful cloning of a new SBP-box gene, *BpSPL1*. Although *BpSPL1* has a similar nucleotide sequence to that of *Antirrhinum SBP2* and *Arabidopsis SPL3*, it does not contain the typical intron sequence that other known *SBP-box* genes have. Northern blot analysis confirmed that *BpSPL1* is expressed in the inflorescence, young stem, and leaves in *B. platyphylla*. An electrophoresis mobility shift assay (EMSA) revealed that a nucleoprotein fragment in *Betula* inflorescence could specifically bind to the SBP-binding element in the *BpMADS* promoter. *BpSPL1* expressed in *Escherichia coli* could also specifically bind to this binding element. This finding suggests that the SBP-box gene is involved in the regulation of floral development in silver birch (Lannenpaa et al. 2004). The expression of the novel *SBP* gene *BpSPL1* in flowers, roots, and leaves was detected in silver birch by Northern blot analysis, further confirming the role of the SBP-box gene in floral development regulation in silver birch (Lännenpää et al. 2004).

8.1.5.3 Roles of SBP Transcription Factors in Leaf Morphogenesis

In recent years, the roles of the *SBP* gene in reproductive growth stage has begun to attract attention from researchers. Overexpression of *AtSPL3* resulted in not only early flowering but also early emergence of hairs on the back of rosette leaves in *Arabidopsis* (Cardon et al. 1999). As with *AtSPL3*, 10 *Arabidopsis SBP* genes are regulated by miR156/157. Overexpression of miR156/157 can significantly increase the number of leaves in young *Arabidopsis*, demonstrating the same effect as downregulation of *SPL3* and other miR156/157 target genes. To further understand the specific role of a single *SBP* gene in the development stage of young *Arabidopsis*, researchers identified and isolated the alleles from single *SBP* mutants. The expression data showed that, compared with other miR156/157-targeted *SPL* genes, *SPL9* and *SPL15* were considerably active in the shoot tips at the vegetative growth stage. Correspondingly, *SPL9* and *SPL15* mutations affected the vegetative growth of the plants. The loss of function of *SPL9* and *SPL15* led to a shortened plastochron, structural changes in the inflorescences, and an increased number of branches in the vegetative growth stage. The morphology of lateral organs in higher plants, such as leaf shape and trichome distribution, changes gradually with branch maturation. In wild-type plants, cauline leaves are oblong with few trichomes, while in transgenic plants with repressed *SPL10/SPL11/SPL2* expression, cauline leaves are wide, and numerous hairs are attached to leaves and flowers. These phenotypes demonstrate a morphology consistent with those of wild-type *SPL10/11/2*-deficient mutants in the reproductive growth stage. This finding suggests that *SPL10*, *SPL11*, and *SPL2* are involved in lateral organ development during branch maturation in the reproductive growth stage. During the vegetative growth stage of *SPL10* transgenic plants, the shape of cauline leaves changes but the trichome distribution remains the same, suggesting that *SPL10* partially regulates branch development in the vegetative growth stage (Schwab et al. 2006).

The wide cauline leaf trait observed in transgenic plants and mutants above is similar to those of the *FUL* mutants. *SPL10* transgenic plants also have an altered expression level of *FRUITFULL*. The SBP protein may play a role in the maturation process of branches by regulating the *Ful* gene (Shikata et al. 2009). Neither *AtSPL9* nor *AtSPL15* loss-of-function mutants show significant developmental changes. However, compared with single gene-deficient mutants, double *AtSPL9/AtSPL15* mutants show a stronger phenotype, affecting the transition from juvenile to adult phase and increasing the leaf plastochron. These findings indicate that *AtSPL9* and *AtSPL15* are functionally redundant and involved in the transition process of vegetative growth (Schwab et al. 2006). *AtSPL14*, a large-sized SBP family member, contains 1035 amino acids. The *Fbr6* (fumonisin B1-resistant) mutant can survive the process of programmed cell death and induces the production of biotoxin fumonisin B1. In addition, in the presence of fumonisin B1, the mutant plants show the phenotype of elongated petioles, enhanced leaf margin serration, and accelerated aging, which can be attributed to the insertion of T-DNA into the *SBP* gene *AtSPL14*. This finding suggests that *AtSPL14* and *FBR6* have similar functions, and both play a role in plant sensitivity to pathogens (fumonisin B1) and regulate the development of normal plant architecture, especially leaf morphology (Stone et al. 2005). Overexpression of *AtSPL3* and its homologous genes *AtSPL4* and *AtSPL5* can lead to *Arabidopsis* phenotypes such as increased leaf epidermal hairs and mature leaves containing smaller but more cells, especially in the first mature leaf (Gandikota et al. 2007a). This finding reveals that *AtSPL3*, *AtSPL4*, and *AtSPL5* are involved in the regulation of vegetative growth in the transition from juvenile to adult phase in *Arabidopsis*. In wild-type *Arabidopsis*, the cauline leaves are oblong, and there are very few epidermal hairs on the leaf surface and flower surface. In contrast, *AtSPL10/ASPL11/AtSPL2*-overexpressing plants have wider cauline leaves and a significantly increased number of epidermal hairs on the leaf surface and flower surface, which is similar to the phenotype in the early reproductive growth stage of wild-type plants. *AtSPL3*-, *AtSPL4*-, and *AtSPL5*-deficient mutants have similar phenotypes, suggesting that these three genes are functionally redundant, and all control lateral organ development by affecting the maturation time of meristem (Shikata et al. 2009).

In maize, the nucleotide sequence change in the promoter and gene-coding region of the *SBP* family gene *TEOSINTE GLUME ARCHITECTURE 1* (*TGA1*) affects the morphogenesis of maize floral meristems (Wang et al. 2005). The study of maize *mop1* and *ts1* lost-of-function mutants has revealed that proper feminization of the male tassel is associated with the increased mRNA expression of several *SBP* gene family members. The maize *SBP* family was expanded from 9 members to 31 members, which are divided into six subfamilies. It was also found that the *SBP* family gene *tasselsheath4* (*tsh4*) is involved in bract development and the establishment of meristem boundaries in maize (Chuck et al. 2007).

8.1.5.4 Other Roles of *SBP* Transcription Factors

A total of 19 *OsSPL* genes have been identified in the *O. sativa* genome. Based on phylogenetic analysis of *SQUA* promoter-binding protein domain, OsSPL proteins are divided into six subclasses. A study conducted by Miura K on *OsSPL* genes investigated the *SQUA* promoter-binding protein domain and identified various exon–intron structure and conserved motifs. In addition, the study analyzed transcriptional expression of *OsSPL* genes in different tissues and organs of *O. sativa* and found that more than half of *OsSPL* genes were significantly expressed in young panicles of rice plants (Miura et al. 2010). In addition, studies have shown that *OsSPL14* and *OsSPL16* play meaningful roles in determining rice quality and yield. Riese et al. isolated and identified 13 *SBP* genes in *Physcomitrella patens* (*P. patens*). Through phylogenetic analysis of conserved SBP-DBD and other molecular characteristics, the researchers established a clear orthologous relationship between *Physcomitrella SBP* genes and the *SBP* family members in flowering plants. The results showed that the *Physcomitrella SBP* genes were concentrated in four groups, of which one is composed of only moss genes and the other three include *Arabidopsis* and *O. sativa SBP* family members. This finding laid a foundation for research on the evolution of SBP transcription factors in various species (Riese et al. 2007). In addition, the study found that SBP transcription factors affect cryptochrome-mediated blue light signal transduction in *P. patens* and promote caulonema side branch formation (Riese et al. 2007).

Only very few studies have been conducted to investigate SBP transcription factors in fruit trees. In tomato (*S. lycopersicum*), 15 SBP members have been identified, and their specific expression in tissues and organs has been analyzed according to their gene structural characteristics. Most tomato *SBP* genes are significantly expressed during the development and maturation of inflorescence and in young fruits. These findings laid a foundation for studying the biological function of SBP transcription factor in berry plants. Further research revealed that the key gene *CNR* (*colorless nonripening locus*) controlling tomato fruit ripening is an *SBP* gene (*LeSPL-CNR*), and methylation modification mutation in its promoter region could inhibit fruit ripening (Fitz-James and Cavalli 2022). A study using epigenetic mutants of *LeSPL*, a member of the *SlSBP* family, confirmed that the *LeSPL* gene is involved in regulating the ripening of tomato fruit. This finding provides strong evidence for the involvement of epigenetic modification in the development and natural variations of tomato and other plants (Manning et al. 2006). To test the hypothesis that *SBP* family genes are also involved in flower induction of *Antirrhinum majus*, researchers used a reverse genetic pathway to silence *AmSBP1* and found that *AmSBP1* could initiate flowering by activating floral meristem identity genes (Preston and Hileman 2013). This finding verified the effectiveness of virus-mediated gene silencing in *Antirrhinum majus*, thereby providing novel insight regarding the role of genetic pathways in protecting flowering time in core dicotyledons.

8.2 Research Progress on LeSPL-CNR

Tomato is a model plant in the research on the development and ripening of fleshy fruit (Arazi and Khedia 2022). Tomato ripening involves significant changes in various physiological and biochemical attributes. The regulatory network formed by transcription factors is a focus of the current research on the molecular mechanisms of fruit ripening. *LeSPL-CNR* is one of the key transcription factors regulating tomato fruit ripening and has two unique characteristics when compared with other fruit-ripening-related transcription factors. First, *LeSPL-CNR* is an epigenetic regulatory gene (Liu et al. 2022). By comparing wild-type *Ailsa Craig* (*AC*) and mutant *Colorless nonripening* (*Cnr*), Manning K et al. found that there was no difference in the nucleotide sequence of *LeSPL-CNR*, but the cytosine in a 286-bp DNA fragment upstream of the *LeSPL-CNR* promoter was highly methylated in the mutant *Cnr*. This finding for the first time demonstrated that epigenetic regulatory factors contribute to the transcriptional regulatory network for fruit ripening (Manning et al. 2006). Second, the 3′-UTR of *LeSPL-CNR* mRNA contains a sly-miR156/157 target site, which has been confirmed by bioinformatics analysis (Moxon et al. 2008). In tomato fruits overexpressing Sly-miR157, the key ripening-related transcription factors, *LeMADS-RIN*, *LeHB1*, *SlAP2a*, and *SlTAGL1*, showed expression profiles similar to that of *LeSPL-CNR*, indicating that *LeSPL-CNR* is likely to be an upstream positive regulator of *LeMADS-RIN*, *LeHB1*, *SlAP2a*, and *SlTAGL1* (Ferreira e Silva et al. 2014).

8.2.1 Discovery and Development of *LeSPL-CNR*

The role of epigenetic factors in tomato fruit ripening was first recognized in the tomato mutant *Cnr* identified in 1993. *Cnr* is characterized by nonripening fruit, a yellow peel, and hard fruit even at the late fruit development stage. In addition, the altered expression of many ripening-related enzymes results in reduced carotenoid synthesis in the peel and other abnormalities in the mutant plants. Compared with wild-type fruits at the same stage, the mutants have smaller amounts and lower activities of polygalacturonase and pectinesterase, which are associated with fruit softening during the late fruit development stage (Eriksson et al. 2004). Site-specific cloning revealed that the *LeSPL-CNR* gene is located on the long arm of the second chromosome in tomato and encodes SBP-box, which contains a highly conserved DNA-binding region (approximately 80 amino acids). Tomato *LeSPL-CNR* belongs to the *SPL* gene family and has the highest similarity with *AtSPL3*. The *LeSPL-CNR* gene is epigenetically regulated, and its nucleic acid sequence does not demonstrate any difference between wild-type *AC* and mutant *Cnr*. However, in *Cnr*, a 286-bp fragment with a particularly high degree of cytosine methylation is located at the upstream of *LeSPL-CNR* promoter; as a result, the DNA methylation change may disrupt the proper ripening of fruit (Manning et al. 2006). In *AC* plants in which fruit *LeSPL-CNR* gene was silenced by virus-induced gene silencing (VIGS) technology, the fruit phenotype was similar to that of *Cnr*, suggesting that *LespL-CNR* plays an important role in tomato fruit ripening.

8.2.2 Biological Functions of *LeSPL-CNR* with *Cnr* Mutant

The phenotype of the tomato *Cnr* mutant, which was identified in 1993, displays significant differences from other known mutants. After 40–50 days of flowering, the fruit remains very hard with yellow skin, and the flesh is colorless or white with low ethylene production. The *Cnr* phenotype cannot be reversed to the wild-type phenotype by exposure to exogenous ethylene. In addition, the cells at the pericarp cannot expand, the intercellular spacing increases, and the cell-to-cell adhesion mediated by homogalacturonan and calcium is weakened at the fruit-ripening stage, leading to altered texture of fruit (Ferreira e Silva et al. 2014). Biochemical tests and gene chip analysis showed that the expression of a large number of ripening-related genes in *Cnr* mutants changed throughout the fruit development period, for example, *RIN* expression significantly increased, and *FUL1* expression significantly decreased, indicating that *CNR*, *RIN*, *NOR*, and *FUL1* at least partially participate in the same transcriptional regulatory pathway (Eriksson et al. 2004).

8.2.3 *LeSPL-CNR* and Epigenetic Regulation

Epigenetic changes are defined as heritable phenotype changes that do not involve alterations in the DNA sequence (Lee and Kim 2022). Their mechanisms include histone modification and DNA methylation, of which DNA methylation is the most important and exists in all higher organisms. DNA methylation has three main forms: 5-methylcytosine, N6-methyladenine, and 7-methylguanine. CCA/TGG and GATC are often methylated in prokaryotes, while only cytosine can be methylated in eukaryotes. In plants, methylation mainly occurs in three forms of cytosine: symmetrical CG and CHG and asymmetric CHH. DNA methylation converts cytosine (5C) at the 5′ end of CpG dinucleotide to 5′methylcytosine (5MeC) under the action of DNA methyltransferases (DNMTs). There are three DNMTs in mammals: DNMT1, DNMT2, and DNMT3 (Del Castillo Falconi et al. 2022). It is generally believed that DNMT3 is involved in the process of DNA remethylation and that DNMT1 is essential for the maintenance of methylated DNA. In plants, DRM2, which is homologous to DNMT3, is involved in the remethylation process. For methylation maintenance, METHYLTRANSFERASE (MET1), a homolog to DNMT1, catalyzes the methylation of the CG site, and CHROMOMETHYLASE (CMT3), a plant-specific DNA methyltransferase, catalyzes the methylation of the CHG site. Methylation of the asymmetric CHH site is catalyzed by DRM2, which may involve some small interfering RNAs (siRNA) or signaling molecules in the process (Jung et al. 2011). DNA methylation regulates gene expression without involving gene sequence alterations; the underlying mechanism DNA methylation can turn off the activity of some genes, and demethylation can induce their reactivation and expression. Therefore, DNA methylation plays a key role in maintaining normal cell function, genetic imprinting, embryonic development, and other biological processes.

Cnr mutant results from spontaneous epigenetic variations, i.e., DNA methylation changes. In *Cnr* mutant plants, a 286-bp fragment upstream of the *LeSPL-CNR* promoter has a particularly high degree of cytosine methylation, and this hypermethylated sequence may repress the expression of *LeSPL-CNR*. A study conducted by Manning K et al. on the tomato spontaneous mutant *Cnr* for the first time revealed the involvement of epigenetic regulatory factors in the transcriptional regulatory network of fruit ripening (Manning et al. 2006). Later, eight methylation-related genes (*SlMET1*, *SlCMT2*, *SlCMT3*, *SlCMT4*, *SlDRM5*, *SlDRM6*, *SlDRM7*, and *SlDRM8*) were identified in tomato. Semiquantitative analysis showed that, despite a spatiotemporal specificity, the expressions of these genes generally decrease with the declining of the degree of DNA methylation in the late fruit-ripening stage, indicating that the degree of methylation decreases at the late fruit-ripening stage. Further, whole-genome methylation sequencing was performed to investigate the regulatory role of DNA methylation in tomato fruit ripening. The sequencing results showed that for most ripening-related genes, the promoter sequences are highly methylated before fruit ripening to prevent the binding of transcription factors, and as fruit develops further, the promoter sequences are gradually demethylated to allow transcription factor recognition and expression. This process indicates that methylation regulation may play an important role in the regulatory effect of transcription factors on some important genes in the early fruit-ripening stage of tomato (Zhong et al. 2013). However, only a few studies have explored the role of epigenetic changes in tomato fruit ripening, and the specific pathways and regulation mechanism involved need to be further studied.

8.2.4 Reevaluated the CNR Function by CRISPR/Cas9 Genome Mutation Technique

Clustered regularly interspaced short palindromic repeats (CRISPR)/ CRISPR-associated protein 9 (Cas9) endonuclease system is a gene-editing technology derived from the adaptive immune system of Streptococcus pyogenes (Wong et al. 2015), which has become a powerful tool to study gene function and the improvement of productivity traits in several plant species (Li et al. 2018).

To further identify the CNR function with CRISPR/Cas 9 in tomato fruit ripening, four specific sgRNAs that targeted the tomato *CNR* were selected by CRISPR-P (http://cbi.hzau.edu.cn/crispr) (Figure 8.1a), and the gRNAs were amplified and cloned into the pYLCRISPR/Cas9Pubi-H binary vector, then was transformed into the wild-type tomato Ailsa Craig using stable *Agrobacterium tumefaciens*. After three generation planting, homozygous CRISPR/Cas9 *CNR*

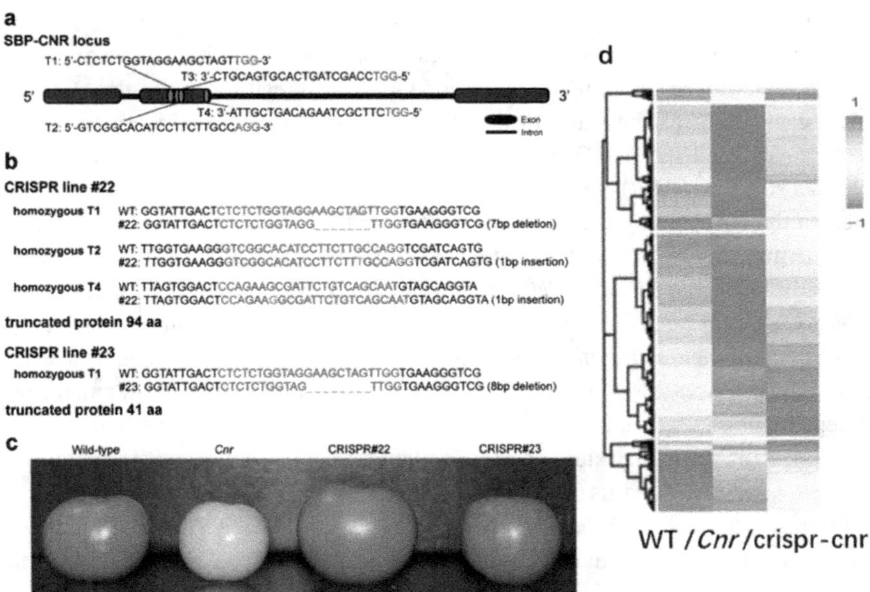

Figure 8.1 Phenotype of the CNR CRISPR/Cas9 knockout. a. Schematic illustration of four gRNA target sites in CNR locus (T1 535–557 bp, T2 564–586 bp, T3 604–582 bp, T4 712–690 bp). b. Sanger sequencing of the CRISPR-edited sites in line #22 (seven bases of AAGCTAG located in 547–553 bp were deleted, one base of T and one base of G were inserted after 580 bp and 695 bp, respectively) and #23 (eight bases of GAAGCTAG located in 546–553 bp were deleted). Red letters indicate the target sites, green letters represent edited sites, and blue letters represent the PAM. c. The CNR CRISPR/Cas9 fruits only showed 2–3 days of delayed ripening compared with that of the WT and the fruits obtained the full color finally, while the original *cnr* mutant failed to ripen. d. Heatmap showing ripening gene expression in the wild-type, *cnr* mutant and CRISPR/Cas9 lines. *Cited from Gao et al. (Gao et al. 2019/Oxford University Press/CC BY 4.0.).*

mutants with Cas9-free were produced (Gao et al. 2019). Sanger sequencing of the CRISPR-edited sites were shown in line #22 (seven bases of AAGCTAG located in 547–553 bp were deleted, one base of T and one base of G were inserted after 580 bp and 695 bp, respectively) and #23 (eight bases of GAAGCTAG located in 546–553 bp were deleted) (Figure 5b). However, the phenotype of CNR knockout fruits is completely different from that of *Cnr* mutants. *CRISPR-cnr line#22* and *line#23* fruits can turn red, while *Cnr* mutant ripening fruits appear yellow (Figure 8.1c). Ripening-related genes expression profile seems also different in *Cnr* mutant and *CRISPR-cnr* ripen fruits by RNA-Seq (Figure 8.1d). The above results from *CRISPR-cnr* fruit indicated that the function of CNR in tomato fruit-ripening control still need be further validated.

8.2.5 Regulatory Networks between Transcription Factors

Tomato fruit development and ripening involves a series of precise transcriptional regulation steps. A variety of genes with important biological functions are usually regulated by multiple transcription factors in either a direct or stepwise manner. In addition, the participating transcription factors can regulate themselves or regulate each other at the transcription level. Three MADS-domain transcription factors—RIPENING INHIBITOR (*RIN*) (Vrebalov et al. 2002), COLORLESS NONRIPENING (*CNR*), and SBP transcription factors (Manning et al. 2006)—as well as the NAC family transcription factors (Martel et al. 2011) can all regulate the ripening-related processes through early transcriptional activation and cascade amplification. *RIN* and *CNR* mutations effectively prevent the ripening process by preventing fruit from producing ethylene and to be unresponsive to exogenous ethylene. This effect suggests that these two genes are located upstream of ethylene biosynthesis and are coregulators that are either dependent on or independent of ethylene. Fruit-ripening traits such as autocatalytic ethylene production, fruit softening, and carotenoid pigment accumulation are inhibited in both *RIN* and *Cnr* mutants (Vrebalov et al. 2002).

Interestingly, *CNR* is the direct target of *RIN* but not *FUL* because *FUL1* expression is considerably low in *Cnr* mutants. The decreased *FUL1* expression in *Cnr* mutants may affect the formation of tetramer complex, and as a result, *CNR* affects the binding of *RIN* to many other target genes. Some ripening regulators such as *RIN* and *CNR* act upstream of *SlAP2* and positively regulate its expression. *CNR* can target and bind to the *SlAP2a* promoter in vitro (Martel et al. 2011). In addition, the mRNA expression of *CNR* increased in the pulp of *SlAP2a* RNAi fruit, indicating that *SlAP2a* and *CNR* are members of the negative feedback regulatory loop; however, the underlying mechanism needs to be further studied. The binding of *LeMADS-RIN* to the target site requires the presence of *LeSPL-CNR* or its regulatory gene products. *SlAP2a* is located downstream of *LeNAC-NOR*, *LeMADS-RIN*, and *LeSPL-CNR*. *LeSPL-CNR* can directly bind to the promoter of *SlAP2a*, while *SlAP2a* can reversely regulate the expression of *LeSPL-CNR* (Martel et al. 2011). Using the bimolecular fluorescence complementation (BiFC) system and yeast two-hybrid screening library, Bemer et al. demonstrated that, while there is no interaction between *SlFUL1* and *SlFUL2*, they both can interact with *LeMADS-RIN* (Bemer et al. 2012). Gel retardation assay showed that *SlFUL2*, *LeMADS-RIN*, and *SlTAGL1* could form a tetramer complex, and *SlFUL1* and *SlFUL2* could each form a complex with *SlTAGL1*, of which *SlFUL2-TAGL2* is the main form (Fujisawa et al. 2011). These studies shed light on the interaction network mechanism among transcription factors related to tomato fruit ripening and help clarify the regulation of downstream structural genes such as ripening-related enzymes and receptor proteins at the levels of transcription, protein structure, and function correlation.

8.3 Prospects

Tomato fruit ripening is a complex genetic process involving phenotypic, physiological, and biochemical changes. The genetic basis of this process has been preliminarily understood through molecular biological analysis of ethylene synthesis and ethylene-mediated signal transduction pathways. The known ripening mutants, including *RIN*, *NOR*, and *Cnr* mutants, have normal fruit size and seed development, but the fruit cannot ripen, does not have an obvious respiratory climacteric, and cannot be reversed to wild-type phenotype by exposure to exogenous ethylene. These characteristics indicate that the fruit-ripening regulatory system in mutants is defective. Considering that the proteins coded by the three genes, *RIN*, *NOR* and *CNR*, are transcription factors, the findings above indicate that transcription factors are a class of important ripening regulatory factors that can function independently or cooperatively with plant hormones. The regulation of transcription factors is very conservative. For example, the RIN-like MADS-box gene, which is a known regulator of floral development and fruit ripening, emerged evolutionarily even earlier than the divergence of monocotyledons and dicotyledons (Giovannoni 2007). As revealed by spectral analysis of transcriptional regulation, a large number of transcription factors are differentially expressed during fruit development and ripening and can regulate these processes either positively or negatively. They act in many aspects, including plant hormone synthesis, cellular swelling, cell wall metabolism, fruit-ripening regulation, and pigment accumulation. However, the existing studies have not sufficiently clarified the regulation mechanism of the transcription factor network and the interactions between transcription factors. For example, *LeMADS-RIN*, *LeNAC-NOR*, and *SlTAGL1* all function in the upstream of the ethylene regulatory pathway. However, exogenous expression of *SlTAGL1* can rescue the phenotypes of *RIN* and *NOR* mutants, while *RIN* and *NOR* mutant fruits have a normal endogenous *SlTAGL1* expression. There is no explanation for this contradictory phenomenon yet (Fujisawa et al. 2013). Hence, further studies are needed to address many unknown factors in the regulatory network involving transcription factors. Future research should focus on (1) continuing to explore and identify ripening-related transcription factors in tomato, determine their downstream targets and mechanisms of action, and then investigate the complete transcriptional regulatory network, which can be verified and compared with other crop fruits; (2) exploring the effects of transcriptional regulation on the metabolic process and quality of tomato fruit and cultivating new varieties of high-quality tomato and other fruit crops by using molecular-genetic approaches for breeding; (3) finding more evidence to support the role of epigenetic modifications in regulating tomato fruit ripening; and (4) clarifying the relationship between fruit ripening and senescence.

References

Arazi, T. and Khedia, J. (2022). Tomato MicroRNAs and their functions. *Int J Mol Sci* 23: 11979.

Arazi, T., Talmor-Neiman, M., Stav, R. et al. (2005). Cloning and characterization of micro-RNAs from moss. *Plant J Cell & Mol Biol* 43: 837–848.

Bartel, D. (2004). MicroRNAs: genomics, biogenesis, mechanism, and function. *Cell* 116: 281–297.

Bemer, M., Karlova, R., Ballester, A.R. et al. (2012). The Tomato FRUITFULL Homologs TDR4/FUL1 and MBP7/FUL2 regulate ethylene-independent aspects of fruit ripening. *Plant Cell* 24: 4437.

Birkenbihl, R.P., Jach, G., Saedler, H., and Huijser, P. (2005). Functional dissection of the plant-specific SBP-domain: overlap of the DNA-binding and nuclear localization domains. *J Mol Biol* 352: 585–596.

Cardon, G., Höhmann, S., Klein, J. et al. (1999). Molecular characterisation of the Arabidopsis SBP-box genes. *Genes Dev* 237: 91–104.

Cardon, G.H., Hohmann, S., Nettesheim, K. et al. (1997). Functional analysis of the Arabidopsis thaliana SBP-box gene SPL3: a novel gene involved in the floral transition. *Plant J* 12: 367–377.

Chuck, G., Meeley, R., Irish, E. et al. (2007). The maize tasselseed4 microRNA controls sex determination and meristem cell fate by targeting Tasselseed6/indeterminate spikelet1. *Nat Genet* 39: 1517–1521.

Del Castillo Falconi, V.M., Torres-Arciga, K., Matus-Ortega, G. et al. (2022). DNA methyltransferases: from evolution to clinical applications. *Int J Mol Sci* 23: 8994.

Din, M.S.u., Wang, X., Alamery, S. et al. (2022). Identification of CT novel polymorphism in 3rd exon of OsSPL14 gene governing seed sequence in rice. *Plos One* 17: e0264478.

Eriksson, M., Moseley, J.L., Tottey, S. et al. (2004). Genetic dissection of nutritional copper signaling in chlamydomonas distinguishes regulatory and target genes. *Genetics* 168: 795–807.

Ferreira e Silva, G.F., Silva, E.M., Azevedo Mda, S. et al. (2014). microRNA156-targeted SPL/SBP box transcription factors regulate tomato ovary and fruit development. *Plant J* 78: 604–618.

Fitz-James, M.H. and Cavalli, G. (2022). Molecular mechanisms of transgenerational epigenetic inheritance. *Nat Rev Genet* 23: 325–341.

Fujisawa, M., Nakano, T., and Ito, Y. (2011). Identification of potential target genes for the tomato fruit-ripening regulator RIN by chromatin immunoprecipitation. *BMC Plant Biol* 11: 26.

Fujisawa, M., Nakano, T., Shima, Y., and Ito, Y. (2013). A large-scale identification of direct targets of the tomato MADS box transcription factor RIPENING INHIBITOR reveals the regulation of fruit ripening. *Plant Cell* 25: 371–386.

Gandikota, M., Birkenbihl, R., Hhmann, S. et al. (2007b). The miRNA156/157 recognition element in the 3′UTR of the Arabidopsis SBP box gene SPL3 prevents early flowering by translational inhibition in seedlings. *Plant J* 49: 683–693.

Gandikota, M., Birkenbihl, R.P., Hhmann, S. et al. (2007a). The miRNA156/157 recognition element in the 3′UTR of the Arabidopsis SBP box gene SPL3 prevents early flowering by translational inhibition in seedlings. *Plant J* 49: 683–693.

Gao, Y., Zhu, N., Zhu, X.F. et al. (2019). Diversity and redundancy of the ripening regulatory networks revealed by the fruitENCODE and the new CRISPR/Cas9 CNR and NOR mutants. *Hortic Res* 6: 39–491.

Giovannoni, J.J. (2007). Fruit ripening mutants yield insights into ripening control. *Curr Opin Plant Biol* 10: 283–289.

Guo, A.Y., Zhu, Q.H., Gu, X.C. et al. (2008). Genome-wide identification and evolutionary analysis of the plant specific SBP-box transcription factor family. *Gene* 418: 1–8.

Guo, J., Song, J., Wang, F., and Zhang, X.S. (2008). Genome-wide identification and expression analysis of rice cell cycle genes. *Plant Mol Biol* 64: 349–360.

Jeena, G.S., Singh, N., and Shikha, Shukla, R.K. (2022). An insight into microRNA biogenesis and its regulatory role in plant secondary metabolism. *Plant Cell Rep* 41: 1651–1671.

Jiao, Y., Wang, Y., Xue, D. et al. (2010). Regulation of OsSPL14 by OsmiR156 defines ideal plant architecture in rice. *Nat Genet* 42: 541–544.

Jung, J.H. and Park, C.M. (2007). MIR166/165 genes exhibit dynamic expression patterns in regulating shoot apical meristem and floral development in Arabidopsis. *Planta* 225: 1327–1338.

Jung, J.H., Seo, P.J., Kang, S.K., and Park, C.M. (2011). miR172 signals are incorporated into the miR156 signaling pathway at the SPL3/4/5 genes in Arabidopsis developmental transitions. *Plant Mol Biol* 76: 35–45.

Kazuhiko Yamasaki, T.K., Inoue, M., Tateno, M., and Yamasaki, T. (2004). A novel Zinc-binding Motif revealed by solution structures of DNA-binding domains of arabidopsis SBP-family transcription factors. *J Mol Biol* 337: 49–63.

Kazuhiko Yamasaki, T.K., Inoue, M., Yamasaki, T., and Yabuki, T. (2006). An Arabidopsis SBP-domain fragment with a disrupted C-terminal zinc-binding site retains its tertiary structure. *FEBS Letters* 580: 2109–2116.

Klein, J. and Huijser, S.H. (1996). A new family of DNA binding proteins includes putative transcriptional regulators of the Antirrhinum majus floral meristem identity gene SQUAMOSA. *Mol. Gen. Genet.* 250: 7–16.

Kropat, J., Tottey, S., Birkenbihl, R.P. et al. (2005). A regulator of nutritional copper signaling in Chlamydomonas is an SBP domain protein that recognizes the GTAC core of copper response element. *Proc Natl Acad Sci USA* 102: 18730–18735.

Lannenpaa, M., Jänönen, I., Holtta-Vuori, M. et al. (2004). A new SBP-box gene BpSPL1 in silver birch (Betula pendula). *Plant Physiol* 120: 491–500.

Lee, J., Park, J., Kim, S.L. et al. (2007). Mutations in the rice LIGULELESS gene result in a complete loss of the auricle, ligule, and laminar joint. *Plant Mol Biol* 65: 487–499.

Lee, J.-S. and Kim, N.-S. (2022). Genomic perspectives on epigenetics. *Genes & Genomics* 44: 247–249.

Li, R., Li, R., Fu, D.Q. et al. (2018). Multiplexed CRISPR/Cas9-mediated metabolic engineering of c-aminobutyric acid levels in Solanum lycopersicum. *Plant Biotechnol J* 16: 415–427.

Liu, Z., Wu, X., Liu, H. et al. (2022). DNA methylation in tomato fruit ripening. *Physiol Plant* 174: e13627.

Manning, K., Tor, M., Poole, M. et al. (2006). A naturally occurring epigenetic mutation in a gene encoding an SBP-box transcription factor inhibits tomato fruit ripening. *Nat Genet* 38: 948–952.

Martel, C., Vrebalov, J., Tafelmeyer, P., and Giovannoni, J.J. (2011). The tomato MADS-box transcription factor RIPENING INHIBITOR interacts with promoters involved in numerous ripening processes in a COLORLESS NONRIPENING-dependent manner. *Plant Physiol* 157: 1568–1579.

Miura, K., Ikeda, M., Matsubara, A. et al. (2010). OsSPL14 promotes panicle branching and higher grain productivity in rice. *Nat Genet* 42: 545–549.

Moreno, M.A., Harper, L.C., Krueger, R.W. et al. (1997). liguleless1 encodes a nuclear-localized protein required for induction of ligules and auricles during maize leaf organogenesis. *Genes Dev* 11: 616–628.

Moxon, S., Jing, R., Szittya, G. et al. (2008). Deep sequencing of tomato short RNAs identifies microRNAs targeting genes involved in fruit ripening. *Genome Res* 18: 1602–1609.

Preston, J.C. and Hileman, L.C. (2013). Functional evolution in the plant SQUAMOSA-PROMOTERBINDINGPROTEIN-LIKE (SPL)genefamily. *Front Plant Sci* 4: 80.

Rhoades, M.W., Reinhart, B., Lim, L.P. et al. (2002). Prediction of plant microRNA targets. *Cell* 110: 513–520.

Riese, M., Hohmann, S., Saedler, H. et al. (2007). Comparative analysis of the SBP-box gene families in P. patens and seed plants. *Gene* 401: 28–37.

Samad, A.F.A., Ali, N.M., Ismail, I., and Murad, A.M.A. (2016). Analysis of miRNAs targeting transcription factors in Persicaria minor induced by Fusarium oxysporum. *AIP Conf Proc* 1784: 020009.

Schwab, R., Ossowski, S., Riester, M. et al. (2006). Highly specific gene silencing by artificial microRNAs in Arabidopsis. *Plant Cell* 18: 1121–1133.

Shao, C.X., Takeda, Y., Hatano, S. et al. (1999). Rice gene encoding the SBP domain protein, which is a new type of transcription factor controlling plant development. *Rice GenetNewsl* 16: 114.

Shikata, M., Koyama, T., Mitsuda, N., and Ohme-Takagi, M. (2009). Arabidopsis SBP-box genes SPL10, SPL11 and SPL2 control morphological change in association with shoot maturation in the reproductive phase. *Plant Cell Physiol* 50: 2133–2145.

Spanudakis, E. and Jackson, S. (2014). The role of microRNAs in the control of flowering time. *J Exp Bot* 65: 365–380.

Stone, J.M., Liang, X.W., Nekl, E.R., and Stiers, J.J. (2005). Arabidopsis AtSPL14, a plant-specific SBP-domain transcription factor, participates in plant development and sensitivity to fumonisin B1. *Plant J* 41: 744–754.

UnteUS, S., Pesaresi, P., Gandikota, M. et al. (2003). SPL8, an SBP-box gene that affects pollen sac development in Arabidopsis. *Plant Cell* 15: 1009–1019.

Vrebalov, J., Ruezinsky, D., Padmanabhan, V. et al. (2002). A MADS-box gene necessary for fruit ripening at the tomato ripening-inhibitor (Rin) locus. *Science* 296: 343–346.

Wang, H., Nussbaum-Wagler, T., Li, B. et al. (2005). The origin of the naked grains of maize. *Nature* 436: 714–719.

Wang, J.W., Czech, B., and Weigel, D. (2009a). miR156-regulated SPL transcription factors define an endogenous flowering pathway in Arabidopsis thaliana. *Cell* 138: 738–749.

Wang, J.W., Schwab, R., Czech, B. et al. (2008). Dual effects of miR156-targeted SPL genes andCYP78A5/KLUH on plastochron length and organ size in Arabidopsis thaliana. *The Plant Cell* 20: 1231–1243.

Wang, Y., Hu, Z., Yang, Y. et al. (2009b). Function annotation of an SBP-box gene in Arabidopsis based on analysis of co-expression networks and promoters. *Int J Mol Sci* 10: 116–132.

Wang, Y., Lan, Q.K., Zhao, X. et al. (2016). Comparative profiling of microRNA expression in soybean seeds from genetically modified plants and their near-isogenic parental lines. *PLoS ONE* 11: e0155896.

Wong, N., Liu, W., and Wang, X. (2015). WU-CRISPR: characteristics of functional guide RNAs for the CRISPR/Cas9 system. *Genome Biol* 16: 218.

Xie, K., Wu, C., and Xiong, L. (2006). Genomic organization, differential expression, and interaction of SQUAMOSA PROMOTER-BINDING-LIKE transcription factors and microRNA156 in rice. *Plant Physiol* 142: 280–293.

Yamasaki, H., Hayashi, M., Fukazawa, M. et al. (2009). SQUAMOSA PROMOTER BINDING PROTEIN-LIKE7 Is a central regulator for copper homeostasis in Arabidopsis. *Plant Cell* 21: 347–361.

Zhang, Y., Schwarz, S., Saedler, H., and Huijser, P. (2007). SPL8, a local regulator in a subset of gibberellin-mediated developmental processes in Arabidopsis. *Plant Mol Biol* 63: 429–439.

Zhong, S., Fei, Z., Chen, Y.-R. et al. (2013). Single-base resolution methylomes of tomato fruit development reveal epigenome modifications associated with ripening. *Nature Biotechnology* 31: 154–159.

Index

Page locators in **bold** indicate tables. Page locators in *italics* indicate figures. This index uses letter-by-letter alphabetization.

t

www.ingramcontent.com/pod-product-compliance
Lightning Source LLC
LaVergne TN
LVHW011522200926
848484LV00001B/107
* 9 7 8 1 3 9 4 1 8 7 6 7 6 *